Ladislav Kvasz

Patterns of Change

Linguistic Innovations in the Development
of Classical Mathematics

Birkhäuser
Basel · Boston · Berlin

Ladislav Kvasz
Comenius University
Faculty of Mathematics, Physics and Informatics
Mlynska dolina
842 48 Bratislava
Slovakia

and

Charles University
Faculty of Education
M. D. Rettigove 4
116 39 Praha
Czech Republic
e-mail: ladislavkvasz@gmail.com

2000 Mathematical Subject Classification: 01-02, 00A30, 00A35, 97C30, 97D20

Library of Congress Control Number: 2008926677

Bibliographic information published by Die Deutsche Bibliothek
Die Deutsche Bibliothek lists this publication in the Deutsche Nationalbibliografie;
detailed bibliographic data is available in the Internet at http://dnb.ddb.de

ISBN 978-3-7643-8839-3 Birkhäuser Verlag AG, Basel - Boston - Berlin

© 2008 Birkhäuser Verlag AG
Basel · Boston · Berlin
P.O. Box 133, CH-4010 Basel, Switzerland
Part of Springer Science+Business Media
Printed on acid-free paper produced from chlorine-free pulp. TCF∞
Cover illustration: Ambrogio Lorenzetti, Annunciation, see page 116.
Printed in Germany

ISBN 978-3-7643-8839-3   e-ISBN 978-3-7643-8840-9

9 8 7 6 5 4 3 2 1   www.birkhauser.ch

For Milena

Acknowledgements

The present book is the result of my research in the field of history and philosophy of mathematics that spanned more than twenty-five years. During this time I was helped by contacts with many experts in the field in my home country, Slovakia, as well as abroad. I would like to express my appreciation and gratitude to all of them – my teachers, colleagues, students or simply friends – who during this rather long period of time encouraged and supported my work. In Slovakia these were my teachers, friends and later colleagues: Peter Cvik, Vladimír Burjan, Milan Hejný, Egon Gál, Pavol Zlatoš, and Peter Bugár. With many of them I met at different private seminars during the communist era. Some became university professors, others never entered academia, but they all were part of an unforgettable atmosphere of parallel culture in the 1980s.

I feel a special debt to Professor Dimitri D. Sokoloff, whom I met during my stays at the Moscow State University in 1989–1991. Thanks to this encounter I enjoyed a glimpse into the scientific culture of the circle round Jakov B. Zeldovich. Thanks to a *Herder scholarship* from the *Stiftung F.V.S. zu Hamburg* I had the privilege of spending the academic year 1993/1994 at the University of Vienna, discussing with Professor Friedrich Stadler, Michael Stöltzner, and Martin Dangl the philosophy of Ludwig Wittgenstein and the Vienna circle. Thanks to a *Masaryk scholarship* from the *Foreign and Commonwealth Office* I spent the academic year 1994/1995 at King's College London studying the work of Karl Popper and Imre Lakatos with Professor Donald Gillies. In 1998 I was awarded a *Fulbright scholarship* and spent a semester at the University of California at Berkeley studying with Professor Paolo Moncosu the role of visual representations in mathematics. And finally, thanks to a Humboldt scholarship, I spent two fruitful years at the Technische Universität Berlin working with Professor Eberhard Knobloch. They all influenced the picture of the development of mathematics presented in this book.

I would especially like to thank Professors Donald Gillies and Graham D. Littler for their help with the final version of the manuscript, to Donald Gillies for writing the preface and to Professor Eberhard Knobloch for his support of the publication of the book.

Some chapters of the present book are based on material published earlier: History of Geometry and the Development of the Form of its Language. *Synthese*, **116** (1998), 141–186; On classification of scientific revolutions. *Journal for General Philosophy of Science*, **30** (1999), 201–232; Changes of Language in the Development of Mathematics. *Philosophia mathematica*, **8** (2000), 47–83; Lakatos' Methodology Between Logic and Dialectic. In: *Appraising Lakatos*. Eds. G. Kampis, L. Kvasz and M. Stöltzner, Kluwer (2002), pp. 211–241. Similarities and differences between the development of geometry and of algebra. In: *Mathematical Reasoning and Heuristics*, (C. Cellucci and D. Gillies, eds.), King's College Publications, London 2005, pp. 25–47; and History of Algebra and the Development of the Form of its Language. *Philosophia Mathematica*, **14** (2006), pp. 287–317. I extend thanks to Kluwer Publishers, Philosophia Mathematica, and Oxford University Press for permission to use these passages.

The financial support of a grant from the Slovak Scientific Grant Agency VEGA 1/3621/06 *Historical and philosophical aspects of exact disciplines* is herewith gratefully acknowledged.

Preface

Kvasz's book is a contribution to the history and philosophy of mathematics, or, as one might say, the historical approach to the philosophy of mathematics. This approach is for mathematics what the history and philosophy of science is for science. Yet the historical approach to the philosophy of science appeared much earlier than the historical approach to the philosophy of mathematics. The first significant work in the history and philosophy of science is perhaps William Whewell's *Philosophy of the Inductive Sciences, founded upon their History.* This was originally published in 1840, a second, enlarged edition appeared in 1847, and the third edition appeared as three separate works published between 1858 and 1860. Ernst Mach's *The Science of Mechanics: A Critical and Historical Account of Its Development* is certainly a work of history and philosophy of science. It first appeared in 1883, and had six further editions in Mach's lifetime (1888, 1897, 1901, 1904, 1908, and 1912). Duhem's *Aim and Structure of Physical Theory* appeared in 1906 and had a second enlarged edition in 1914. So we can say that history and philosophy of science was a well-established field by the end of the 19th and the beginning of the 20th century.

By contrast the first significant work in the history and philosophy of mathematics is Lakatos's *Proofs and Refutations*, which was published as a series of papers in the years 1963 and 1964. Given this late appearance of history and philosophy of mathematics relative to history and philosophy of science, we would expect the early development of history and philosophy of mathematics to be strongly influenced by ideas which had been formulated and discussed by those working in the history and philosophy of science. This proves to be the case. Lakatos's own pioneering work was, as the title indicated, developed from Popper's model of conjectures and refutations which Popper had devised to explain the growth of science.

In 1992, I edited a collection of papers with the general title: *Revolutions in Mathematics*. Once again the title showed clearly that ideas drawn from the history and philosophy of science were being applied to mathematics. The publication of Kuhn's 1962 *The Structure of Scientific Revolutions* led to debates about whether revolutions occurred in science and, if so, what was their nature. The 1992 collection carried over this debate to mathematics.

Kvasz's book on history and philosophy of mathematics breaks to some extent with this tradition of importing ideas from the history and philosophy of science into mathematics. This is because he adopts a *linguistic approach*. Kvasz's idea (p. 6)[1] is to interpret the development of mathematics as a sequence of linguistic innovations. Kvasz points out (p. 7) several advantages of this approach. These include the fact that languages have many objective aspects that can easily be studied, and so are more accessible to analysis than, for example, heuristics, or psychological acts of discovery.

As a result of his linguistic approach, Kvasz draws more on ideas from general analytic philosophy than from philosophy of science. More specifically he makes use of the classic works of Frege and the early Wittgenstein. However, Kvasz develops the ideas of Frege and the early Wittgenstein in a number of novel ways. Perhaps most importantly he introduces a historical dimension to the study of language. Both Frege and Wittgenstein treat language as timeless, but Kvasz, by contrast, focusses on the historical changes by which an older language can develop into a stronger, richer new language which has greater expressive power.

Perhaps, however, Kvasz has not broken away completely from philosophy of science, because it is worth noting that Kuhn too adopted a linguistic approach in his later period. This is shown in the 2000 book *The Road Since Structure* which contains essays by Kuhn from the period 1970–1993. On p. 57 of this book, Kuhn goes as far as to say:

---

[1]   Page references on their own are to Kvasz's book of which this is the preface.

"If I were now rewriting *The Structure of Scientific Revolutions*, I would emphasize language change more and the
normal/revolutionary distinction less." (Kuhn)

This passage comes from a talk which Kuhn gave in 1982, twenty
years after the publication of *Structure*.

Yet although Kuhn adopted a linguistic approach, he did not man-
age to produce results in his new research programme at all comparable
in significance to those of his earlier period. By contrast, as we shall
see, Kvasz does indeed produce a number of novel and exciting results.
I will argue later that one reason for Kvasz's greater success using the
linguistic approach is that it is more suitable for the analysis of mathematics than for the analysis of science, though the linguistic approach
does still have some value for science.

Kvasz, using his linguistic approach, formulates three patterns of
change in the development of mathematics. These are: (1) *re-codings*,
(2) *relativizations*, and (3) *re-formulations*. I will now briefly describe
these in turn.

A re-coding occurs (p. 8) when there are changes in the formation
rules of terms and formulae, or in the rules for construction of geometrical figures. Kvasz takes a quotation from Frege's 1891 paper *Funktion und Begriff* as the starting point for his analysis of re-coding. In
this quotation (given on p. 15), Frege traces a development which starts
with simple arithmetical assertions, such as $2 + 3 = 5$, goes on to general algebraic laws, such as $(a + b).c = a.c + b.c$, and then to the
coinage of the technical term "function", and the statement of general
laws about functions. Each step in this development is, according to
Frege, the transition to a higher level.

Kvasz develops this idea in a number of significant ways. First of
all he points out that Frege concentrates on symbolic languages, which
are those dealing with arithmetic, algebra and analysis. Kvasz suggests
that we consider, on a par with such symbolic languages, what he calls
iconic languages, in a similar fashion, with geometrical figures. In this
approach (pp. 12–13) geometrical diagrams are not just heuristic aids,
but an integral part of the geometrical theory.

The introduction of iconic languages leads to one of Kvasz's most
interesting claims which is that the path of development does not go
directly from one symbolic language to another as the Frege quotations
seem to suggest, but rather via an iconic intermediate level. So, according to Kvasz, there is not a direct transition from elementary arithmetic
to algebra. Rather the transition is first from elementary arithmetic to

synthetic geometry. This was carried out by the Greeks. Next there is a transition from synthetic geometry to algebra which was carried out later by the Arabs. Indeed according to Kvasz, this oscillation continues, since algebra leads to the iconic language of analytic geometry which in turn leads to the symbolic language of differential and integral calculus. A diagram showing all these transitions is to be found on p. 86.

Here it can be remarked that, although Kvasz's main focus is on linguistic change, he does often mention wider cultural factors which may have influenced mathematical development. Thus he says (pp. 29–30):

> "It seems that there must have been some obstacle that prevented the Greeks from entering the sphere of algebraic thought. The first who entered this new land were the Arabs. There is no doubt that they learned from the Greeks what is a proof, what is a definition, what is an axiom. But the Arabic culture was very different from the Greek one. Its center was Islam, a religion which denied that transcendence could be grounded in the metaphor of sight. Therefore the close connection between knowledge and sight which formed the core of the Greek *epistéme*, was lost." (Kvasz)

Kvasz's next pattern of change is relativization (pp. 8–9). This differs from re-coding (pp. 7–8) in that the ways of generating descriptions remain unchanged, but there are changes in the relation between the linguistic expressions and the objects that they stand for. Kvasz takes some ideas from Wittgenstein's *Tractatus* as the starting point of his analysis of relativization. In the *Tractatus*, Wittgenstein presents the picture theory of language. Regarding pictures, he makes the following important observation:

> "2.172 A picture cannot, however, depict its pictorial form:
> it displays it." (Wittgenstein)

On the picture theory of language, then, language must have a form which is displayed in the language, but cannot be expressed in the language. To Kvasz, this suggests a way in which an initial language L_1 say can be transformed historically into a more powerful language L_2. We consider the form of the language of L_1 which cannot be expressed within L_1 itself. However by adding this form to L_1 we create a new language L_2 which has more expressive power than L_1. The creation of more powerful languages in this manner does indeed, according to Kvasz, occur frequently in the development of mathematics. It is what he calls relativization.

Kvasz first example of relativization gives another example of his cultural leanings. He considers the system of perspective used by Renaissance painters. This was of course mathematical in character and its language had a form which Kvasz calls the perspectivist form. Perspective paintings are designed to be seen from a particular viewpoint, but this viewpoint lies outside the painting. It is something which cannot be expressed in the perspectivist form of language. Suppose, however, that we incorporate this point (renamed the centre of projection) into our system. We then get projective geometry whose language goes beyond that of perspectivism.

This is a striking and convincing example of what Kvasz calls 'relativization'. However, it is far from the only one. He goes on to analyse, using this concept, the emergence of ever more complicated forms of synthetic geometry – including most notably non-Euclidean geometry. This approach casts new light on some of the puzzling features of the discovery of non-Euclidean geometry.

The pattern of relativization seems to be naturally suited to developments in geometry, but Kvasz's next surprising claim is that it applies to algebra as well. The sequence of successive relativizations in algebra is not quite the same as the sequence in geometry, as can be seen by comparing the diagrams on pp. 160 and 200. However there is still a great deal of similarity. Thus while with recodings we have an oscillation between the symbolic and the iconic, with relativizations the symbolic and the iconic develop independently but in parallel, passing through most of the same stages.

Kvasz's third pattern of change, re-formulation, introduces, as the name suggests, less dramatic changes than the other two. However, re-formulations can still be of great importance in bringing about advances in mathematics. Kvasz first examples of re-formulations are taken from Lakatos's *Proofs and Refutations*, and this leads him in Chapter 4 to give an interesting analysis of Lakatos's work. Kvasz claims that the changes which Lakatos considers, both in mathematics and science, are all re-formulations, and so relatively small in character. Lakatos never deals with the bigger changes which occur in re-codings or relativizations. This seems to me quite plausible. Lakatos has the unique honour of having been the first to introduce the historical approach into philosophy of mathematics. However, being a pioneer in this respect, it was

likely that he would only succeed in dealing with one type of change –
leaving others to be discovered later.

One impressive feature of Kvasz's book is the wealth of detailed
historical case studies from the history of mathematics with which he
illustrates his three patterns of change. Now there might be some arguments as to whether one or other of the examples given are really examples of the pattern which they are supposed to exemplify. However,
these are likely to be disagreements on the margin, and there seems to
me little doubt that the three patterns described by Kvasz are indeed
patterns which have characteristically recurred in the development of
mathematics. The concepts of recoding and relativization are quite
novel and original, and so must be considered as constituting a real
and substantial contribution to history and philosophy of mathematics.
Both concepts go beyond anything to be found in Lakatos. They show
that Kvasz has made genuine progress with his linguistic approach, and
in this respect has done better than Kuhn whose linguistic approach did
not lead to any such striking results. One reason for this situation, in my
view, is that mathematics is more suited than science to the linguistic
approach. I will now explain why I think this to be the case.

Let us compare a big innovation in science such as the development
of special relativity with a big innovation in mathematics such as the
development of calculus. The development of calculus brought a lot of
linguistic changes. There were the introduction of signs for differentials such as dy/dx, and for integrals, such as $\int$. The results of the new
calculus were expressed in formulae which were quite different from
those of previous mathematics and would have been as incomprehensible to earlier mathematicians as hieroglyphics were to Egyptians living
in the 18th Century. The introduction of the calculus was a recoding in
Kvasz's sense. When we turn now to the introduction of special relativity, this brought about enormous conceptual changes in science. Ein-
steinian mass is quite different from Newtonian mass. Yet the language
used to express the new results of special relativity differed hardly at
all from the immediately preceding language of Newtonian mechanics. This example, and others like it, indicated that mathematics is the
place to look for those who want to study how significant changes in
language occur. The language of every day life does indeed change,
but at a slow rate which can take centuries. In a few decades mathematicians can develop strikingly new languages. Thus mathematics is
a kind of artifical laboratory in which large linguistic changes can be
observed and studied with ease.

The importance of mathematics for the study of linguistic change shows that Kvasz's results and the various schemas which he introduces such as recoding and relativization should be of interest not just to philosophers of mathematics, but also to those with a general interest in the philosophy of language. Perhaps some of the results obtained by Kvasz will have application in other fields. Indeed Kvasz himself considers some examples from the history of art, and notions such as 'iconic language' could well have a fruitful application in the field of aesthetics.

There is however a problem because the book presupposes a considerable knowledge of mathematics and its history which could well render its results unintelligible to someone without this background. The way round this problem for philosophers of language who are not mathematicians would perhaps be to read a selection from the book which gives the main ideas without too many technical details. A suitable such selection might be the following: Introduction (pp. 1–10), Ch. 1 to the end of 1.1.3 (pp. 11–37) + first section of 1.2 (pp. 85–89), Ch. 2 to the end of of 2.1.2 (pp. 107–124) + 2.2 to the end of 2.2.2 (pp. 160–172), Ch. 3 to the end of 3.1 (pp. 225–232), Ch. 4 (pp. 239–251). Even for the more technically minded reader, it might be worth reading these sections first to get an overview of Kvasz's system, before plunging into the details of the examples from the history of mathematics which Kvasz uses to support his ideas.

So to sum up. Kvasz has made a considerable advance in the field of history and philosophy of mathematics by adopting a linguistic approach. This has enabled him to formulate three patterns of change which are at the same time patterns of linguistic innovation. At least two of these (recoding and relativization) are quite novel and nothing like them has so far been discussed. His work therefore deserves careful study by anyone interested in history and philosophy of mathemetics. Moreover, because the patterns of linguistic innovation may perhaps also be found in other areas where new linguistic forms develop, Kvasz's work also deserves the attention of those with more general interests outside mathematics in the philosophy of language.

University College London, March 2008   *Donald Gillies*

Contents

Introduction

The question of the nature of change in the development of mathematics was systematically discussed for the first time in the second half of the nineteenth century in connection with the discovery of non-Euclidean geometries. This discussion was evidence of a fundamental change in the perception of mathematics. Even at the beginning of the eighteenth century it was still common for mathematicians to try to give legitimacy to, and to underline the importance of, their discoveries by ascribing them to ancient authorities. In doing so they implicitly assumed that mathematical theorems express eternal and permanent truths and thus a discovery is only an incidental event when someone becomes aware of these eternal truths. This strategy of ascribing his own discoveries to ancient authors was used even by Newton. According to the testimony of Nicolas Facio de Drivillier, Newton was convinced that all important theorems of his *Philosophiae Naturalis Principia Mathematica* were known already to Plato and Pythagoras (Rattansi 1993, p. 239).

The idea of *progress* that emerged during the era of Enlightenment and the image of a *creative genius* that was influential during Romanticism changed in a radical way this view of the nature of discoveries in mathematics. In the course of discussions about the discovery of the new geometries, which started in the second half of the nineteenth century, the personality of the author stood in the foreground. The context in which the question of the nature of change in the development of mathematics was discussed was thus the *context of psychology of discovery*. This was rather natural as most of the authors participating in this discussion, such as Hermann von Helmholtz, Eugenio Beltrami, Felix Klein, or Henri Poincaré, were themselves active mathematicians who had a personal, one could say, intimate relation to the question of change in mathematics. In their texts they tried to articulate this relation. In this connection perhaps the most well-known passage is a passage from Poincaré's book *Science et Méthode*:

> "For a fortnight I had been attempting to prove that there could not be any functions analogous to what I have since called Fuchsian functions. I was at that time very ignorant. Every day I sat down at my table and spent an hour or two trying a great number of combinations, and I arrived at no result. One night I took some black coffee, contrary to my custom, and was unable to sleep. A host of ideas kept surging in my head; I could almost feel them jostling one another, until two of them coalesced, so to speak, to form a stable combination. When morning came, I had established the existence of one class of Fuchsian functions, those that are derived from the hypergeometric series. I had only to verify the results, which only took a few hours.
>
> Then I wished to represent these functions by the quotient of two series. This idea was perfectly conscious and deliberate; I was guided by the analogy with elliptical functions. I asked myself what must be the properties of these series, if they existed, and I succeeded without difficulty in forming the series that I have called Theta-Fuchsian." (Poincaré 1908, pp. 52–53)

In the quoted passage Poincaré analyzed the interplay of intuition and purposeful thought in mathematical creativity. This and similar texts determined the tone of the discussion in the philosophy of mathematics around the end of the nineteenth century. Scientists who were, during the creative period of their lives, protagonists of fundamental changes in mathematics now, in their later years, attempted a philosophical reflection about these changes.

After the discovery of the logical paradoxes in the early years of the twentieth century the interest of mathematicians shifted gradually towards the foundations of mathematics. Gottlob Frege, David Hilbert, Giuseppe Peano, and Bertrand Russell could be mentioned as the initiators of the debate on the foundations of mathematics. This debate continues to the present day – see Schirn 1998 or Shapiro 2005. One of the main distinctions introduced during the debate on the foundations of mathematics was the distinction between the context of discovery and the context of justification. According to this distinction, in connection with each mathematical proposition, two fundamentally different sets of questions can be asked. The questions of the first set inquire who, when, and under what circumstances *discovered* the particular proposition. These questions belong to psychology, sociology, or history and so by asking them we find ourselves after a short time outside of philosophy of mathematics. The questions of the second set ask under what conditions is the proposition true, what suppositions are necessary for its *justification*, which propositions it contradicts. Asking these questions we remain in the philosophy of mathematics proper.

From the point of view of the distinction between the context of discovery and the context of justification, the debate on the foundation of mathematics belonged unequivocally to the context of justification. On the other hand, the question of the nature of change in mathematics, which dominated the philosophy of mathematics at the end of the nineteenth century, belonged to the context of discovery. As the proponents of the mentioned distinction were convinced that philosophy of mathematics should be confined to the context of justification, efforts were made to exclude the question of the nature of change in mathematics from the philosophy of mathematics altogether and to relocate it in the field of psychology. The psychological context in which this question was discussed at the end of the nineteenth and the beginning of the twentieth century only fuelled these efforts at relocation. Although the efforts to exclude from philosophy of mathematics the question of the nature of change in mathematics did not succeed, the *criticism of psychologism* put forward by proponents of the foundationalist approach drew attention to the necessity of an intersubjective reformulation of this question.

Therefore when, due to the works of George Polya *How to solve it?* (Polya 1945), *Mathematics and Plausible Reasoning* (Polya 1954), and *Mathematical Discovery* (Polya 1962) as well as to a series of papers by Imre Lakatos *Proofs and Refutations* (Lakatos 1963–64), the discussion on the nature of change in mathematics came back, the context of psychology of discovery was replaced by the *context of methodology of change*. The problems of correctness of a proof, of plausibility of an argument, of reliability of a heuristics or of definiteness of a refutation came into the center of attention. All these problems concern the intersubjective aspect of a discovery. In this context Polya developed his *methods of plausible reasoning* and Lakatos formulated his *method of proofs and refutations*. The turn from psychology of discovery to the methodology of change proved to be a lucky one, because due to this turn philosophy of mathematics found itself making contact with discussions in the philosophy of science.

When in 1962 Thomas Kuhn published *The Structure of Scientific Revolutions*, an interesting dialogue started between the philosophy of mathematics and the philosophy of science. On the one side Lakatos, by adapting his views on the philosophy of mathematics, proposed the *methodology of scientific research programs* (Lakatos 1970) and entered into a polemic with Kuhn in the field of philosophy of science. On the other side, as the question of the nature of change in mathematics was transferred into the methodological context, it became possible to adopt Kuhn's position also to mathematics and to formulate the problem of the possibility of revolutions in mathematics. In this connection we can mention the papers of Michael Crowe *Ten "laws" concerning patterns of change in the history of mathematics* (Crowe 1975), Herbert Mehrtens *T. S. Kuhn's theories and mathematics: a discussion paper on the "new historiography" of mathematics* (Mehrtens 1976) and Joseph Dauben *Conceptual revolutions and the history of mathematics: two studies in the growth of knowledge* (Dauben 1984). These papers initiated a vivid discussion that was summarized in the collection *Revolutions in Mathematics* (Gillies 1992). This indicates that the context of methodology is rather fruitful for discussion of the nature of change in the development of mathematics and has continued to attract attention up to the present – see *The Growth of Mathematical Knowledge* (Grosholtz and Breger 2000).

During the 1950s and 1960s, when discussion of the nature of change in the development of mathematics was dominated by methodological questions, an important change took place in the field of analytic philosophy. Under the influence of the work of Ludwig Wittgenstein and authors such as John Austin, Donald Davidson, Paul Grice, Jaakko Hintikka, Hilary Putnam, Willard van Orman Quine, and Wilfried Sellars, the way in which questions were formulated and answered changed fundamentally. As an illustration of this radical shift in the view of the role of language in philosophy of mathematics we can mention the discussion of Kant's philosophy of geometry in the works of Jaakko Hintikka and Michael Friedman. In his reconstruction of the role of intuition in Kant's philosophy of geometry, Hintikka challenged the generally accepted view that dominated in the analytic tradition since Hilbert's *Grundlagen der Geometrie* (Hilbert 1899). According to this view, shared by such philosophers as Bertrand Russell or Rudolph Carnap, geometrical figures are only psychological aids which can enhance our understanding of a particular theory but contribute nothing to the theory's content. In contrast to this, Hintikka argued that figures and constructions in intuition play in Euclid's *Elements*, which was the paradigmatic example of a mathematical theory for Kant, an important role analogous to existential instantiation in modern logic (Hintikka 1965, p. 130). Thus geometrical figures are not merely didactical or psychological tools but are constitutive parts of the logical structure of geometrical theories. This argument was further developed by Friedman who has shown that until modern quantification theory was developed (i.e., until the publication of Frege's *Begriffsschrift* in 1879) geometrical figures had to be used in order to compensate for the weakness of the so-called monadic logic (for details see Friedman 1985, p. 466–468).

The arguments of Hintikka and Friedman are important because they view the language of mathematics not as some ideal form existing outside of time, but rather understand the language of mathematics as a historically changing system. In this way they introduce a *historical dimension* into the foundations of mathematics. Mathematics is always done by dint of some linguistic tools and the historicity of these tools bestows a historical dimension to the foundations of mathematics. Nonetheless, this model of historicity is rather different than the one that is usually understood. The historicity introduced into the foundations of mathematics by Hintikka and Friedman is not caused by some external influence (social, political, or cultural). The historicity of the foundations of mathematics is internal; it is predicated upon the historicity of the language of mathematics. It is due to the fact that the logical and expressive tools which we use in building a mathematical theory are historical. At particular moments in history some logical derivations were simply technically impossible.

Another important aspect of the arguments of Hintikka and Friedman is that they understand *geometrical figures* as an important component of the language of mathematics which is on a par with its symbolic formulae. It seems that, for the first time in the analytic tradition, geometric figures are viewed not as a mere psychological aid, but as an important tool used in construction of the logical structure of mathematical theories. This idea is of fundamental importance for us, because it offers a theoretical vindication of the approach to the history of geometry employed in the present book. We will base our reconstruction of the development of mathematics on an analysis of the changes in the use and interpretation of geometrical figures that can be found in mathematical texts. Thus, in a sense, the reconstruction of the history of geometry which will follow is a development of the possibilities opened up by Hintikka when he changed our view of geometrical figures.

We have seen how, at the beginning of the twentieth century, criticism from the foundationalist camp in the philosophy of mathematics gave an important impulse for a reformulation of the question of the nature of change in mathematics. The criticism caused a transition from discussion of this question in the *context of psychology of discovery* to analyzing it in the *context of methodology of change*. It seems that a shift in the analytic tradition, illustrated by the ideas of Hintikka and Friedman, can give a similar impulse for a shift in formulation of the question of the nature of change in mathematics. It is possible that the time has come to replace the *context of methodology of change*, i.e., the context that was used by Polya and Lakatos in their analysis of mathematical discovery, heuristics, or refutations by a new context, the *context of linguistic change*. This context draws our attention to problems such as: what *linguistic innovations* were accompanying a particular mathematical discovery; what the linguistic framework of a particular heuristics was; what *linguistic means* were used in a particular refutation. It turns out that the fundamental discoveries in the history of mathematics were closely connected with important linguistic innovations. The great discoverers were as a rule also great linguistic innovators. And often it was the change of language, the change of the rules of syntax or semantics, that enabled a mathematician to express connections that were until then inexpressible, and so arrive at a discovery. When we shift the discussion of changes in the development of mathematics from the context of methodology to the context of language, the whole discussion of changes in mathematics obtains much firmer foundations.

Of course, the question of language was a central topic in discussions of the foundations of mathematics since Frege, Peano, and Hilbert. Nevertheless, the problem was that in these discussions the language of mathematics was viewed as an ideal, atemporal logical calculus. If we stop viewing mathematics on the background of an ideal logical language and accept the historicity of the linguistic tools of mathematics, it becomes possible to analyze the changes of these tools in the development of mathematics. Thus we can *interpret the development of mathematics as a sequence of linguistic innovations.* In developing this approach to the history of mathematics we will follow three principles:

1. Instead of the *context of psychology of discovery* or the *context of methodology of change* we will analyze the nature of change in mathematics in the context of its language. Thus we will interpret changes in mathematics as changes of the *language* of mathematics.

2. We will try to understand the language of mathematics historically, i.e., not as some ideal calculus that exists outside of time. Instead we will try to reconstruct the formal languages of a particular period in history and to understand the changes of language which occurred during the transition from one historical period to the next. Thus we will interpret changes in mathematics as *changes* of the language of mathematics.

3. We will understand the language of mathematics as close to its practice as possible. We will not force upon mathematics any *a priori* understanding of language. We will include in language of mathematics the rules for construction and interpretation of geometrical figures as well as rules of formation and use of symbolic expressions. Thus we will interpret changes in mathematics as changes of the language of *mathematics.*

The main advantage of the linguistic approach to the reconstruction of changes in the development of mathematics lies in the fact that it enables us to discriminate several kinds of change. As long as we analyze the changes in mathematics in the framework of psychology of discovery or in the framework of methodology of change, all changes appear basically the same. The differences that we encounter are due either to differences in the individual psychological make-up of the particular mathematician or in his/her individual methodological preferences, both of which are rather elusive and difficult to study. On the other hand, if we approach a particular change in mathematics from the linguistic point of view, we have at our disposal two languages – the language before and the language after the change. These languages are better accessible to analysis than the psychological act of the discovery or the heuristics that led to it. A language has many objective aspects that can be studied and analyzed. But even more important is that the linguistic approach opens up the possibility of *comparing different changes and of classifying them.* From our analysis of a number of linguistic changes in the history of mathematics, we realized that it is possible to discriminate three levels on which linguistic innovations in mathematics occur.

The first kind of linguistic change concerns *how descriptions are generated.* In the case of symbolic languages these are changes of the formation rules of terms and formulae; in the case of geometrical languages these are changes of the rules of construction of geometric figures. On this level the linguistic innovations unite into a developmental line the languages of *arithmetic, algebra, differential and integral calculus,* and *predicate calculus.* We obtain this developmental line if we restrict ourselves to the symbolic side of mathematics and follow the main syntactic innovations that occurred in history (such as the introduction of variables in algebra, or the introduction of functions in the differential and integral calculus). In a similar way we can obtain a developmental line connecting *synthetic geometry, analytic geometry,* and *fractal geometry.* We obtain this developmental line if we restrict ourselves to the geometric side of mathematics and follow the main innovations in the way that geometric objects are constructed. I suggest calling these changes of language of mathematics **re-codings** and they are discussed in the first chapter of the book.[1]

The second kind of linguistic change leaves the ways of generation of descriptions untouched but it changes the *relation between the linguistic expressions and the objects that they stand for.* On this level the linguistic innovations unite into a developmental line *Euclid's Elements, projective geometry, Lobachevski's non-Euclidean geometry, Klein's Erlanger Program,* and *algebraic topology.* Along this line the rules for the construction of linguistic descriptions did not change. All these theories use the same ingredients – points, straight lines, and circles – in constructing their objects, and so they all belong to synthetic geometry (in contrast to analytic or fractal geometry, where the ways

---

[1] The present book is an integration of several independent strands of research. In each of them (Kvasz 1998, Kvasz 2000, and Kvasz 2006) I tried to find appropriate terminology for the different patterns of change in the development of mathematics. As each of the mentioned papers dealt with only one pattern, it contained only one piece of this new terminology. Now, when I am trying to unite the pieces together I decided to change the terminology so that it will be closer to standard English. Thus I will use from now on the term *re-coding* instead of *re-presentation*; and *relativization* instead of *objectification*. Nevertheless, this is just a simple replacement and so I hope it will not cause any problems to a reader who is perhaps familiar with my previous work.

of construction of objects (or rather of their linguistic descriptions) are radically different). What is changing in the above mentioned developmental line is the way that we interpret these descriptions, how we assign meaning to them. A triangle in Euclid, in projective geometry, in Lobachevski, in Klein, or in Poincaré looks the same; it is constructed following the same rules. Nevertheless, in each of these cases it is something rather different because it has different properties and different propositions can be proven about it. The changes of this kind in the language of mathematics I suggest calling **relativizations** and they are discussed in the second chapter of the book.

The third kind of change leaves untouched both the ways of generation of linguistic descriptions, and the relation between the descriptions and the objects that they stand for. It concerns the *different ways of formulating a theory*. As an example we can take the developmental line uniting the different *axiomatizations of Euclidean geometry from Euclid till Hilbert*. Here the universe of geometrical objects did not change (in contrast to a *re-coding* as for instance the transition from synthetic to analytic geometry where many new curves were introduced). Nor was there change in the interpretation of linguistic expressions (in contrast to a *relativization* as for instance Klein's *Erlanger Program* where geometric objects were interpreted as invariants of particular transformation groups). The objects described by the theories of this developmental line and the theorems that can be proven about them, are the same. Nevertheless, what is changing is the formulation of fundamental notions, the list of basic principles, and the overall plan of construction of the theory. I suggest calling changes of this kind *re-formulations* and they are discussed in the third chapter of the book.

The distinction between these three kinds of linguistic change makes it possible to discover remarkable regularities in the development of mathematics. These regularities have remained hidden until now because they defy being analyzed in the framework of methodology of change. In this framework the three kinds of change look all the same – they use similar heuristics and similar strategies of plausible reasoning. Therefore if the analysis of the nature of change in mathematics is undertaken in the framework of methodology of change, specific features of the different kinds of change are lost and what remains is a robust structure common to all three of them. If we choose a particular variant of the methodological approach, it will describe this robust structure from its particular point of view. Thus for instance Kuhn's theory describes the robust structure common to the different kinds of change in mathematics in sociological terms (such as scientific community, paradigm, anomaly, etc.). But neither of the different methodological approaches is able to discover that there are three fundamentally different kinds of change. The first one connects Euclid with Descartes, the second one connects Euclid with Lobachevski and the third one connects Euclid with Playfair.

Only if we focus our attention on the language by means of which a theory is formulated, it becomes possible to discriminate three kinds of change. If we separate them from each other and analyze each of them without the interference of the other two, remarkable regularities appear. Kuhn's theory blurs these regularities in the development of science or mathematics in a similar way as a photographer would blur the contours of a face, if he were to project pictures of three different faces onto the same paper. What would remain on the resulting photograph are the coarse features common to all faces, but all the individual details would be lost. Similarly Kuhn retained only the coarse social dynamics of adaptation of the scientific community to change, the dynamics that is common for all three kinds of change discussed in this book. But the details that are peculiar to *re-codings*, to *relativizations*, and to *re-formulations* are lost in Kuhn's theory. Therefore the analysis of patterns of change in the development of mathematics, that are discussed in this book can be seen as a refinement of Kuhn's theory applied to mathematics.

CHAPTER 1

Re-coding as the First Pattern of Change in Mathematics

The roles of geometry and of arithmetic in contemporary philosophy of mathematics are rather asymmetric. While arithmetic plays a central role in foundational approaches and therefore its logical structure is thoroughly studied and well understood (see Shapiro 2005), geometry is the central topic of the antifoundational approaches (see Boi, Flament, and Salanskis 1992). This of course does not mean that there are no foundational studies of geometry; it is sufficient to mention the work of Alfred Tarski (see Tarski 1948 and 1959). Nevertheless, in these cases geometry is just another illustration of the methods developed for the analysis of arithmetic. The visual aspect of geometry, the very fact that geometry has something to do with space and spatial intuition, is totally ignored in these studies. These studies are just exceptions, and they do not change the basic difference that the philosophy of arithmetic is dominated by the foundational approach, while the philosophy of geometry is mainly antifoundational.

The reason for this asymmetry lies in the different attitudes to the languages of these two main parts of mathematics. Since the works of Frege, Peano, and Russell, the language of arithmetic has been fully formalized, and so the formulas of arithmetic are considered to be a constitutive part of the theory. On the other hand, in geometry the geometrical pictures are considered only as heuristic aids, which can help us to understand the theory but strictly speaking do not belong to the theory itself. Since Hilbert, the content of a geometrical theory is independent of any pictures and is determined by its axioms. Thus in the formalization of arithmetic the specific symbols like "+" or "≤", as well as the rules which they obey, were considered part of the language. In geometry the process of formalization took rather the opposite direction: all the special symbols, like "." or "—", were excluded from the language. An interesting analysis of the reasons for this exclusion of diagrams and of diagrammatic reasoning from the foundations of mathematics is given in (Graves 1997).

Although there were good reasons for such a development of the foundations of mathematics, we believe it might be interesting to try to bridge the gap between the philosophy of arithmetic and the philosophy of geometry. For this purpose it is necessary to do in geometry what Frege did in arithmetic. This means, first of all, formalizing its language and thus turning the pictures from mere heuristic aids into integral parts of the theories themselves. A picture is not just the physical object formed by spots of graphite on the more or less smooth surface of the paper. We understand the picture as an expression (a term) of the iconic language with its own meaning and reference. We follow here an analogy with arithmetic or algebra, where a formula is understood not as a physical object, i.e., not as spots of ink on a sheet of paper.[1] If we succeeded in the incorporation of pictures into the language of mathematics it would enable us to deepen our understanding of those periods of the history of mathematics where geometric pictures and argumentation based upon them played a crucial role (i.e., practically the whole history of geometry before Pasch and Hilbert).

For our purposes it is enough to give a short characterization of the iconic language of geometry. We interpret a picture as a term of the iconic language. Then a geometrical construction becomes a generating sequence of the resulting expression (picture). In this way the

---

[1]   The aim of interpreting geometrical figures as a language may be challenged by posing the problem of how to represent propositions. Nevertheless, it is important to realize that also in the case of symbolic languages the representation of propositions was not introduced overnight. Viète at the end of the sixteenth century, i.e., more than a century after Regiomontanus, did not have a symbol for the expression of identity and so he was unable to express a proposition solely by symbols. Therefore he connected fragments of his symbolic language into propositions by means of ordinary language. Thus also in the case of the language of geometry it is not necessary in a single move to solve all the problems of its syntax. In the following we will restrict ourselves to the representation of terms of the iconic language. Maybe later someone will find a way of forming propositions from terms of the iconic language similarly to the way in which, by means of the symbol of identity, we form propositions from the terms of a symbolic language.

Euclidean postulates ("*To draw a straight line from any point to any point.*" or "*To describe a circle with any centre and distance.*") become formation rules of this language, analogous to the Fregean rules for symbolic languages, which prescribe, how from an n-ary functional symbol F and n terms $t_1, t_2, \ldots, t_n$, a new term $F(t_1, t_2, \ldots, t_n)$ is formed. A picture is called a "well-formed expression" if each construction step is performed in accordance with the formation rules (axioms). If we rewrite the Euclidean postulates 1, 2, 3, and 5 (see Euclid, p. 154) as formation rules, we obtain a general description of the language.[2] The questions when two terms are equal or how we can introduce predicates into the pictorial language, and what its propositions look like, are subtle questions, which we don't want to raise now. They would require more detailed investigations, which would lead us rather far from the subject of our book. For our present purposes it is sufficient to realize that the iconic language of geometry can be treated with the same strength and precision as that which Frege introduced for arithmetic. Seen from this position, mathematics for us will consist no longer of an exact symbolic language supplemented by some heuristic pictures, but rather of two languages of the same rank, one of them symbolic and the other iconic.

Of course the pictures of Euclidean (synthetic) geometry are not the only pictures used in geometry. There are pictures also in analytic (algebraic and differential) geometry as well as in iterative (fractal) geometry. If we complement the language of mathematics by pictures of synthetic, analytic, and fractal geometry, it will in a radical way increase the capacity of our linguistic approach to the study of changes in the development of mathematics. We will describe an interesting periodic motion in the history of mathematics, consisting in alternation of its symbolic and geometrical periods. Thus in ancient Egypt and Babylonia the symbolic approach was dominant. Later, in Ancient Greece, geometry came to the fore and dominated mathematics until the sixteenth century, when a revival of the symbolic approach took place in

---

[2] From this point of view Pasch's discovery, that Euclidean geometry contains no postulate that would guarantee that two circles drawn from the opposite ends of a given straight line with radii equal to the length of that line will intersect, could be interpreted as the discovery of the fact that the corresponding figure is not a well-formed term of the language of synthetic geometry. We, of course, see the points of intersection, but if we realize that a plane containing only points with rational co-ordinates is a model of Euclid's postulates, it becomes obvious that the existence of the points of intersection cannot be proven from the postulates. This example illustrates what it means to view figures as terms of the language of synthetic geometry and to interpret Euclid's axioms as formation rules of this language.

algebra. In the next century, due to the discovery of analytic geometry, the iconic language came again to the fore, while the nineteenth century witnessed a return of the dominance of the symbolic language in the form of the arithmetization of analysis. This phenomenon has not yet been sufficiently understood. Nobody has tried to undertake a serious analysis of it. The reason could be that the alternation of symbolic and geometrical periods in mathematics has a vague and obscure nature, which dissuades people from undertaking its serious analysis. Nevertheless, I am convinced that this vagueness and obscurity is only a result of insufficient understanding of the logical and epistemological aspects of geometrical pictures. As long as pictures are considered vague and obscure, their alternation with formal languages, which is clearly visible in the history of mathematics, must remain vague and obscure as well. By interpreting pictures as iconic language, we create a framework which makes it possible to understand the relations between the symbolic and iconic periods in the history of mathematics. In their regular alterations we will discover the first pattern of changes in the development of mathematics.

1.1. Historical Description of Re-codings

In reconstruction of the linguistic innovations in the history of mathematics that we labeled re-codings we will follow Gottlob Frege. Frege published in 1879 his *Begriffsschrift, eine der arithmetischen nachgebildete Formelsprache des reinen Denkens* in which he presented the modern quantification theory and an axiomatic system of the predicate calculus. In the closing chapter of his book Frege introduced definitions of several notions of the theory of infinite series. He devoted the next fourteen years to further elaboration of ideas from this chapter and in 1893 he published the first volume of his opus magnum: *Grundgesetze der Arithmetik, Begriffsschriftlich abgeleitet.* If we compare the logical framework of these books, we find several important changes. The *Begriffsschrift* is based on a syntactic approach to logic – Frege chose a particular system of formulas which served as axioms and derived from them the whole system of the predicate calculus. The *Grundgesetze* contained several semantic extensions of the logical system of the *Begriffsschrift,* among which perhaps the most important was the introduction of *concepts* as functions whose possible values are *the True* and *the False,* and the introduction of *extensions of concepts* as course-of-values of these functions. Frege's thoughts motivating these changes can be found in the papers published between 1891 and 1892: *Funktion und Begriff, Über Begriff und Gegendstand* and *Über Sinn und Bedeutung.*

The first of these papers contains an idea which we will take as the starting point of our reconstruction of re-codings in mathematics. In his paper Frege described the evolution of the symbolic language of mathematics from elementary arithmetic through algebra and mathematical analysis to predicate calculus:

> " If we look back from here over the development of arithmetic, we discern an advance from level to level. At first people did calculations with individual numbers, 1, 3, etc.
>
> $$2 + 3 = 5 \qquad\qquad 2 \cdot 3 = 6$$
>
> are theorems of this sort. *Then they went on* to more general laws that hold good for all numbers. What corresponds to this in symbolism is the transition to the literal notation. A theorem of this sort is
>
> $$(a + b) \cdot c = a \cdot c + b \cdot c .$$
>
> At this stage they had got to the point of dealing with individual functions; but were not yet using the word, in its mathematical sense, and had not yet formed the conception of what it now stands for. *The next higher level* was the recognition of general laws about functions, accompanied by the coinage of the technical term 'function'. What corresponds to this in symbolism is the introduction of letters like f, F, to indicate functions indefinitely. A theorem of this sort is
>
> $$\frac{\mathrm{d}F(x) \cdot f(x)}{\mathrm{d}x} = F(x) \cdot \frac{\mathrm{d}f(x)}{\mathrm{d}x} + f(x) \cdot \frac{\mathrm{d}F(x)}{\mathrm{d}x} .$$
>
> Now at this point people had particular second-level functions, but lacked the conception of what we have called second-level functions. By forming that, we make *the next step forward.*" (Frege 1891, p. 30; English translation p. 40)

Our interpretation of this development of the symbolic language will differ from Frege's in two respects. The first is terminological: we will not subsume algebra or mathematical analysis under the term "arithmetic" but will rather consider them as independent languages. More important, however, is the fact that we will show how this "development of arithmetic" described by Frege interplayed with geometrical intuition. In order to achieve this we need to complement Frege's analysis of the "development of arithmetic" with a similar "development of geometry". Frege identified the main events in the development of the symbolic language as the introduction of the concept of "individual functions" in algebra, then of the "particular second-level functions" in the calculus and finally the general concept of "second-level functions" of the predicate calculus. In a similar way we will try to identify the crucial events of the development of the iconic language of geometry. It will turn out that the events parallel to those described by Frege are the creation of analytic geometry, fractal geometry and set theory.

The unification of the symbolic and iconic languages enables us to consider the development of mathematics as the evolution of its language. We will study the development of the language of mathematics from the following six aspects:

1. *Logical power* – how complex formulas can be proven in the language,

2. *Expressive power* – what new things can the language express, which were inexpressible in the previous stages,

3. *Explanatory power* – how the language can explain the failures which occurred in the previous stages,

4. *Integrative power* – what sort of unity and order the language enables us to conceive there, where we perceived just unrelated particular cases in the previous stages,

5. *Logical boundaries* – marked by occurrences of unexpected paradoxical expressions,

6. *Expressive boundaries* – marked by failures of the language to describe some complex situations.

The evolution of the language of mathematics consists in the growth of its logical and expressive power – the later stages of development of the language make it possible to prove more theorems and to describe a wider range of phenomena. The explanatory and the integrative power of the language also gradually increases – the later stages of development of the language enable deeper understanding of its methods and offer a more unified view of its subject. To overcome the logical and expressive boundaries, more and more sophisticated and subtle techniques are developed. We will illustrate the growth of the logical, expressive, explanatory, and integrative power, as well as the shifts of the logical and expressive boundaries of the language of mathematics by some suitable examples.[3]

Mathematics has a tendency to improve its languages by addition. So, for instance, we are used to introducing the concept of variable into the language of arithmetic (enabling us to write equations in this language) and often we choose the field of real numbers as a base (so that the language is closed with respect to limits). This is very convenient from the pragmatic point of view, because it offers us a strong language in which we can move freely without any constraints. But, on the other hand, it makes us insensitive to historically existing languages. The old languages do not appear to us as independent systems with their own logical and expressive powers. They appear only as fragments of our powerful language. Since our aim is the epistemological analysis of the language of mathematics, we try to characterize every language as closely as possible to the level on which it was created. We ignore later emendations which consist of incorporation of achievements of the later development into the former languages (for instance of the concept of variable into arithmetic). In this book the language of arithmetic will be a language without variables. We think that such a stratification of the language of mathematics into different historical layers will be interesting also for logical investigations, showing the order in which different logical tools appeared.

1.1.1. Elementary Arithmetic

Counting is as old as mankind. In every known language there are special words expressing at least the first few numbers. For instance the Australian tribes around Cooper bay call 1 – *guna*, 2 – *barkula*, 3 – *barkula guna*, and 4 – *barkula barkula* (Kolman 1961, p. 15). With the development of society it became necessary to count greater quantities of goods and so different aids were introduced in counting: fingers, pebbles, or strings with knots. A remarkable tool was found in Moravia

---

[3] The logical and expressive boundaries of a particular language can be expressed only by means of a later language; a language that is strong enough to enable us to construct a situation by the description of which the original language fails. Therefore for characterization of the logical and expressive boundaries of a particular language we will use later languages.

(Ifrah 1981, p. 111). It was used by our ancestors in the Paleolithic era (19 000 – 12 000 BC) probably for counting bigger quantities. It is a bone of a young wolf on which 55 cuts are visible. The first 25 cuts are rendered in groups by five and the 25th cut is twice as long as the others. This rendering shows traces of quinary notation. Further development of civilization and the related necessity to count bigger and bigger quantities led to a fundamental discovery, which changed mankind – the invention of numerals as special symbols designed for counting. Practically every civilization created its own numeral system, a system that made it possible to reduce counting to manipulation with symbols. We know many different ways in which this reduction can be achieved. Some civilizations took as the basis of their numeral systems 60, others took 10; some civilizations introduced a positional notation, others used a non-positional one. An overview of the principles of different numeral systems can be found in the book of George Ifrah (Ifrah 1981). What all numeral systems have in common is the creation of the first *symbolic language* in history, the language of elementary arithmetic.

The language of elementary arithmetic is the simplest symbolic language. It is based on manipulations with numerical symbols. There are many variants of this language; the most common contains ten symbols for numerals $0, 1, \ldots, 9$ and the symbols $+, -, x, :$, and $=$. A basic feature of this language is that it has no symbol for a variable. For this reason it is impossible in this language to express any general statement or write a general formula. The rules for division or for multiplication, as general statements, are inexpressible in this language. They cannot be *expressed* in the language, but only *shown*. For instance the rule that multiplication by 10 consists in writing a 0 at the end of the multiplied number cannot be expressed in the language. It can only be shown on specific examples such as $17 \times 10 = 170$, or $327 \times 10 = 3270$. From such examples one understands that the particular numbers are unimportant and one grasps the universal rule.

1.1.1.1.   *Logical Power – Verification of Singular Statements*

Typical statements of elementary arithmetic[4] are singular statements such as:

$$135 + 37 = 172 \qquad \text{or} \qquad 24 \times 8 = 192.$$

The language of elementary arithmetic contains implicit rules, which make possible verification of such statements with the help of manipulation with symbols, i.e., on a purely syntactical basis. One consequence of the fact that the language of elementary arithmetic did not contain variables was the necessity of formulating all problems with concrete numbers. Let us take for example a problem from the *Rhind papyrus:*

> "Find the volume of a cylindrical granary of diameter 10 and height 10.
>
> Take away 1/9 of 10, namely 1 1/9; the remainder is 8 2/3 1/6 1/18. Multiply 8 2/3 1/6 1/18 times 8 2/3 1/6 1/18; it makes 79 1/108 1/324 times 10; it makes 790 1/18 1/27 1/54 1/81 cubed cubits. Add 1/2 of it to it; it makes 1185 1/6 1/54, its contents in khar. 1/20 of this is 59 1/4 1/108. 59 1/4 1/108 times 100 hekat of grain will go into it." (Fauvel and Gray 1987, p. 18)

Instead of rewriting the problem as an equation and solving it in a general way, as we would do today, the scribe has to take the numbers from the formulation and to perform with them particular arithmetic operations, until he gets what he needs. The general method, which is inexpressible in the language, is shown with the help of concrete calculations. The language of elementary arithmetic was the first formal language in history, which made it possible to solve problems by manipulation with symbols. Its logical power is restricted to verification of particular statements.

---

4   In his paper *Funktion und Begriff* Frege mentioned as a typical proposition of elementary arithmetic the identity $2 + 3 = 5$. In his *Grundlagen der Arithmetik* Frege criticized Kant and as an example on which it is obvious that the propositions of arithmetic cannot be founded on intuition he mentions $135664 + 37863 = 173527$ (Frege 1884, p. 17). This example clearly shows that the language of elementary arithmetic is based on rules for formal manipulation with symbols.

*1.1.1.2.   Expressive Power – Ability to Express Arbitrarily
         Large Numbers*

The language of elementary arithmetic is able to express arbitrarily large natural numbers. This may seem nothing special, as we are used to negative, irrational and complex numbers and so the natural numbers seem to us as a rather poor and limited system, where it is not possible to subtract or divide without constraints. Nevertheless, if we leave out of consideration these results of later developments, maybe we will be able to feel the fascination which must have seized the ancient Egyptian (Babylonian, Indian, Chinese) scribe when he realized, that *with the help of numbers it is possible to count everything*. This is the basis of bureaucratic planning, which was one of the most important discoveries of the ancient cultures. The universality of bureaucracy is based on expressive power of the language of arithmetic. Traces of fascination with the expressive power of the language of arithmetic can be found in an affinity for large or special numbers in mythology, in the Kabbala, in the Pythagoreans and even in Archimedes. His book *The Sand-reckoner* is devoted to a demonstration of the expressive power of the language of elementary arithmetic. Archimedes shows that numbers are able to count even the grains of sand on the earth:

> "There are some, King Gelon, who think that the number of the sand is infinite in multitude; and I mean by the sand not only that which exists about Syracuse and the rest of Sicily but also that which is found in every region whether inhabited or uninhabited. Again there are some who, without regarding it as infinite, yet think that no number has been named which is great enough to exceed its multitude. And it is clear that they who hold this view, if they imagined a mass made up of sand in other respects as large as the mass of the earth, including in it all the seas and the hollows of the earth filled up to a height equal to that of the highest of the mountains, would be many times further still from recognizing that any number could be expressed which exceeded the multitude of the sand so taken. But I will try to show you *by means of geometrical proofs*, which you will be able to follow, that, of the numbers named by me and given in the work which I sent to Zeuxippus, some exceed not only the number of the mass of sand equal in magnitude to the earth filled up in the way described, but also that

of a mass equal in magnitude to the universe. ...I say then that, even if a sphere were made up of the sand, as great as Aristarchus supposes the sphere of the fixed stars to be, I shall still prove that, of the numbers named in the *Principles* [lost work of Archimedes], some exceed in multitude the number of the sand which is equal in magnitude to the sphere referred to ....." (Archimedes 1952, pp. 520–521 and Heath 1921, pp. 81–85)

Of course, the geometrical proofs by means of which Archimedes shows that it is possible to "name" a number that is greater than the quantity of sand in the whole universe, do not belong to arithmetic. But the very fact that such numbers can be named illustrates the expressive power of the language of elementary arithmetic. The Egyptian or Babylonian scribes were unable to prove such a proposition, but they felt that everything can be counted and thus taken into records.

1.1.1.3. Explanatory Power – The Language of Elementary Arithmetic is Nonexplanatory

The language of arithmetic is *nonexplanatory*. This is obvious from preserved mathematical texts, which have the form of collections of recipes, comprising a sequence of instructions without any explanation. This characteristic of the language has been noticed also by historians. For instance, Gray speaks about "contradictory and nonexplanatory results" (Gray 1979, p. 3).

1.1.1.4. Integrative Power – The Language of Elementary Arithmetic is Nonintegrative

It is also obvious that the language of elementary arithmetic is *nonintegrative*. It does not make it possible to conceive any kind of unity or order. Mathematical texts are mere collections of unrelated particular cases. Ordering of problems is based on their content (problems on calculation of areas of fields, volumes of granaries, etc.) instead of on their mathematical form. This means that the ordering principle comes from outside the language.

1.1.1.5. Logical Boundaries – Existence of Insoluble Problems

The old Babylonians could encounter the fact that some problems have no solution (Gray 1979, p. 24). On Babylonian tables we can find a problem that leads (in modern notation) to the following system of equations

$$x + y = 10, \qquad\qquad x \cdot y = 16,$$

which has two solutions: (2, 8) and (8, 2). The Babylonians were able to solve it in the framework of elementary arithmetic. Nevertheless, if we take a problem slightly different from the previous one:

$$x + y = 10, \qquad\qquad x \cdot y = 40,$$

the Babylonian procedure collapses. In the framework of elementary arithmetic it is impossible to see what has happened. A procedure that works perfectly in some circumstances turns out to be useless if we only slightly change the conditions.

1.1.1.6.   *Expressive Boundaries – Incommensurability of the Side and the Diagonal of a Square*

The Pythagoreans developed a qualitatively new kind of formal language. It was the iconic language of synthetic geometry. Nevertheless, at the beginning they connected this new geometrical language with an interesting kind of "arithmetical atomism". The Pythagoreans supposed that every quantity, among others also the side and the diagonal of a square, comprise a finite number of units. So proportion of the lengths of the side and the diagonal of the square equals the proportion between the numbers of units from which they are composed. The discovery of the incommensurability of the side and the diagonal of the square refuted the Pythagorean atomism.[5] It shows, however, that the language of geometry is more general than that of arithmetic. In arithmetic the side and diagonal of a square cannot be included in one calculation. We can either choose a unit commensurable with the side, but then it will be impossible to express the length of the diagonal by

---

[5] The Pythagoreans believed that every quantity can be expressed by means of natural numbers. In the case of length measurements they measured the length of a particular segment using an appropriately chosen unit. Then the length of the segment could be expressed in the form of a proportion of the segment and the unit. The Pythagoreans believed that each such proportion has the form of the ratio of two natural numbers. In this case they called the segment and the unit commensurable. One of the most fundamental discoveries of ancient Greek mathematics was the discovery that in the case of the square there is no unit that would be commensurable with both its side and its diagonal. There is an extensive literature dealing with this rather famous discovery (see Heath 1921, pp. 90–91; Gray 1979, pp. 11–13; Dauben 1984, pp. 52–57; or Boyer and Merzbach 1989, pp. 79–81).

a number, or we can choose a unit commensurable with the diagonal, but then we will be unable to express the length of the side. So the incommensurability of the side and diagonal of the square reveals the boundaries of the expressive power of the language of elementary arithmetic.[6]

1.1.2. Synthetic Geometry

Since the earliest stages of their development the great agricultural civilizations of antiquity were confronted with several problems that are closely linked to geometry. By measuring fields, planning irrigation channels, constructing roads, building granaries and fortification walls, as well as many other practical activities, they collected knowledge about different geometrical objects. Even the term *"geometry"*, which has its origin in the Greek words for earth (*geos*) and measure (*metros*), reveals the practical roots of this discipline. However, despite the quantity of practical knowledge collected in these ancient civilizations, geometry did not become an independent discipline. It was embedded into the framework of arithmetical calculative recipes just like calculation of the volume of the cylindrical granary, quoted above from the *Rhind papyrus*. Even though the problem is a geometrical one, it is treated in a purely arithmetical way. Geometry as an independent mathematical discipline was created in ancient Greece. Thales and Pythagoras, the founders of geometry, made according to ancient tradition long trips to Egypt and to Mesopotamia where they learned the practical geometrical knowledge of these civilizations. They brought this knowledge back to their homeland and presumably enriched it with many new discoveries that bear their names. But what is even more important, they created a new mathematical language suitable for representing geometrical objects.

---

[6] This limitation was later removed when mathematicians introduced real numbers as their basic number system. But we are interested here in the language of elementary arithmetic in its original form. And for the kind of elementary arithmetic that takes number to be a collection of units, the incommensurability of the side and diagonal of the square is a paradox that defies comprehension. Later this paradox was resolved by the introduction of real numbers and it was replaced by the positive fact that $\sqrt{2}$ is an irrational number. Here we see one of the fundamental strategies of dealing with paradoxes. It consists in the extension of the language and in the reinterpretation of the paradox so that it becomes a positive fact. Nevertheless, it took many centuries until the incommensurability was resolved inside of the symbolic language itself.

Geometrical language, in contrast to the language of arithmetic, is an iconic and not a symbolic language. Its expressions are pictures formed of points, line segments and circles, rather than formulas formed of linear sequences of arithmetic symbols. From this point of view the Greeks moved from the symbolic language of arithmetic to the iconic language of geometry. They had good reasons for this, because, as we will see, geometrical language surpasses the language of elementary arithmetic in logical as well as in expressive power. It makes it possible to prove general theorems and to represent incommensurable line segments such as the side and the diagonal of a square.

1.1.2.1.   Logical Power – Ability to Prove Universal Theorems

The new language of geometry was originally developed in close connection with arithmetic. We have in mind the famous Pythagorean theory of figurate numbers (see Boyer and Merzbach 1989, p. 62). Certainly, its content is arithmetical – it deals with numbers. But its form is quite new. Using small dots in sand or pebbles (*psefos*) it represents numbers geometrically – as square numbers (i.e., numbers the *psefoi* of which can be arranged into a square, like 4, 9, 16, . . .), triangular numbers (like 3, 6, 10, . . .), and so on. With the help of this geometrical form, arithmetical predicates can be visualized. For instance an even number is a number the *psefoi* of which can be ordered into a double row. So the *arithmetical* property of being odd or even becomes a property expressible in the new *geometrical* language. This very fact, that arithmetic properties became expressible in the language makes it possible to prove universal theorems (and not only particular statements, as was the case until then). For instance the theorem that the sum of two even numbers is even can be easily proved using this Pythagorean language. It follows from the fact that if we connect two double rows, one to the end of the other, we will again get a double row. Therefore the sum of *any two* even numbers must be even. We see that the Pythagoreans introduced an important linguistic innovation into mathematics. With the help of their figurate numbers they were able to prove general mathematical theorems.

The discovery of incommensurability led the Greeks to abandon the Pythagorean arithmetic basis of their new geometrical language and to separate geometrical forms from the arithmetic content. In this process the new language, the iconic language of geometry, was created. From the logical point of view this new geometrical language is much stronger than the language of elementary arithmetic. It makes it possible to prove universal statements. The language of geometry is able to do this, thanks to an expression of a new kind – a *segment of indefinite length*. (In fact this was the essence of the Pythagorean innovation – they were able to prove that the sum of any even numbers is even, because the double row which represented an even number could be of any length. The geometrical form is independent of the particular arithmetical value to which it is applied.) If we prove some statement for such a segment, in fact we have proved the statement for a segment of any length, which means that we have proved a general proposition.

The segment of indefinite length is not a variable, because it is an expression of the iconic and not of the symbolic language. That means it does not refer to, but rather represents the particular objects (side of a triangle, radius of a circle, etc.). Of course, any concrete segment drawn in the picture has a precise length, but this length is not used in the proof, which means that the particular length is irrelevant. This substantiates the interpretation of geometrical pictures as a language. In a proof we ascribe to a segment that has a precise length a universal meaning, thus we treat it as a linguistic expression.

1.1.2.2.   *Expressive Power – Ability to Overcome Incommensurability*

The language of geometry is superior to that of elementary arithmetic also regarding its expressive power. For the language of arithmetic the incommensurability of the side and the diagonal of a square means that there are lengths which cannot be expressed by numbers. From the geometrical point of view, the side and diagonal of the square are ordinary segments. Thus the iconic language of synthetic geometry has expressive superiority over the language of elementary arithmetic.

The ratio of the lengths of two incommensurable line segments cannot be expressed by numbers. Nevertheless, the language of geometry makes it possible to compare such lengths. Eudoxus created for this purpose his theory of proportions that can be found in Book V. of Euclid's *Elements*. This theory is based on the following definition:

> "Magnitudes are said to be *in the same ratio*, the first to the second and the third to the fourth, when, if any equi-multiples whatever be taken of the first and third, and any equimultiples whatever of the second and forth, the former equimultiples alike exceed, are alike equal to, or alike fall short of, the later equimultiples respectively taken in corresponding order." (Euclid V., Def. 5)

This definition makes it possible to compare two pairs of incommensurable magnitudes and show that for instance the areas of two "circles are to one another as the squares on the diameters" (Euclid XII, Prop. 2). The areas of a circle and of the square on its diameter are incommensurable magnitudes (as we know today) but despite this incommensurability the language of geometry makes it possible to prove that their ratio is constant for all circles. This illustrates the superiority of the expressive power of the language of synthetic geometry over the language of elementary arithmetic, which was unable to say anything about two incommensurable magnitudes.

1.1.2.3. Explanatory Power – Ability to Explain the Non-Existence of a Solution of a Problem

As an illustration of the logical boundaries of the language of elementary arithmetic we showed that from the arithmetical point of view it is incomprehensible why some problems, as for instance

$$x + y = 10, \qquad\qquad x \cdot y = 40$$

have no solution. The language of synthetic geometry can explain this strange fact:

> "One advantage of the appeal to geometry can now be mentioned to illustrate the gain in *explanatory power*. Evidently there are no numbers x and y whose sum is 10 and product 40, and the Babylonian scribes seem to have avoided discussing such questions. However, we can now see why there are no such numbers. In the language of application of areas we have to put a rectangle of area 40 on a segment of the length 10 leaving a square behind.

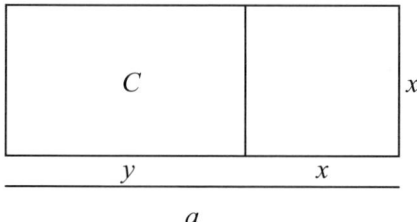

> The area xy of the large rectangle C varies with x (and therefore with y) but is greatest when the rectangle is a square. In that case $x = y = a/2$, and the area is $a^2/4$.

Therefore we can solve the problem provided $a^2/4$ exceeds the specified area C. In our example $100/4 = 25$ is not greater than 40, so no numbers can be found. The discussion of the feasibility of finding a solution is indeed to be found in Euclid (Book VI, Prop. 27) immediately preceding the solution of the quadratics themselves." (Gray 1979, p. 24)

Thus the language of synthetic geometry enables one to understand the conditions under which problems of elementary arithmetic have solutions. In this way the language of geometry makes it possible to express explicitly the boundaries of the language of arithmetic; the boundaries that in the language of elementary arithmetic itself are inexpressible.

1.1.2.4.   *Integrative Power – The Unity of Euclid's Elements*

So far we have shown that the language of synthetic geometry has logical, expressive and explanatory superiority over the language of elementary arithmetic. A further advantage of this language is that it makes it possible to create a unifying approach to mathematics. Perhaps the best illustration of this unifying approach is Euclid's *Elements*. Euclid united number theory, plain geometry, theory of proportions, and geometry in three dimensions into one organic whole. Euclid's *Elements* are only partially original. Substantial parts of it were taken from older sources that are now lost. The *theory of proportions* that comprises Book V of the *Elements* was taken probably from Eudoxus. From him stems also the *method of exhaustion* that can be found in Book XII. The *classification of irrational magnitudes* contained in Book X was taken from Theaetetus and the number theory contained in Books VII.–IX. from Archytas. Euclid's original contribution was probably the theory of the five Platonic solids contained in Book XIII. Thus the *Elements* are a compilation that contains ideas of several mathematicians. Despite their heterogeneous content they form a unity which captures the reader by its stringent logical structure. And this structural unity of the Elements can be seen as the best illustration of the integrative power of the language of synthetic geometry.

1.1.2.5.   *Logical Boundaries – Existence of Insoluble Problems*

There are three problems, formulated during the early development of Greek geometry, which turned out to be insoluble using just the elementary methods of ruler-and-compasses construction. These problems are: *to trisect an angle, to duplicate a cube,* and to *construct a square with the same area as a circle.* The insolubility of these problems was proved with modern algebra and complex analysis, that is, in languages of higher expressive and explanatory power than that of the language of synthetic geometry. From the point of view of synthetic geometry the fact that nobody managed to solve these problems, despite the efforts of the best mathematicians, must have been a paradox. It indicated the logical boundaries of the language of synthetic geometry.

Many solutions of these problems were presented using curves or methods which have been characterized by some historians of mathematics as *ad hoc* (Gray 1979, p. 16). As an example we can take the *trisectrix*, a curve invented by Hippias, or the method of *neusis*, invented by Archimedes (see Boyer and Merzbach, 1989, pp. 79 and 151). Nevertheless, these new curves did not belong to the language of synthetic geometry, as characterized in the introduction. They contained points which could not be constructed using only ruler and compass but were given with the help of other mechanical devices. So even if these new methods are important from the historical perspective and have attracted the attention of historians (see Knorr 1986), they fall outside the scope of our analysis.

1.1.2.6.   *Expressive Boundaries – Equations of Higher Degrees*

The new geometrical language prevailed in Greek mathematics to such a degree that Euclid, when confronted with the problem of solving a quadratic equation, presented the solution as a geometrical construction. It was natural, because in the geometrical setting it was not necessary to care whether the magnitudes involved in the problem were commensurable or not.

> "The discovery of incommensurability and the impossibility of expressing the proportion of any two segments as a proportion of two natural numbers led the Greeks to start to use proportions between geometrical magnitudes instead of arithmetical proportions and with their help to express general proportions between magnitudes.... In order to solve the equation $cx = b^2$, the Greeks regarded the b^2 as a given area, c as a given line segment and x the unknown segment. They transformed thus the problem into the construction of an oblong whose area and one side are known. Or as they

called it, 'the attachment of the area to a given line segment.'." (Kolman 1961, p. 115)

Solving equations with the help of geometrical constructions avoids
incommensurability. However, such methods are appropriate only for
linear and quadratic equations. Cubic equations represented real technical complications, for they deal with volumes. Equations of higher
degree were beyond the expressive boundaries of the language of synthetic geometry.

1.1.3. Algebra

Algebra is a creation of the Arabs. The ancient Greeks did not develop the algebraic way of thinking. This does not mean that they
could not solve mathematical problems which we today call algebraic.
Such problems were solved already in ancient Egypt and Babylonia.
Nevertheless, the Egyptian and Babylonian mathematicians were en-
chanted by their calculative recipes and did not see the need to develop
any more general methods. On the other hand the ancient Greeks excluded calculative recipes from mathematics and reduced almost the
whole of mathematics to geometry. Therefore they lost contact with
symbolic thinking and calculative manipulations. The reduction of algebraic problems to geometry has, moreover, one fundamental disadvantage. The second power of the unknown is in geometry represented
by a square, the third power by a cube; but for higher powers of the unknown there is no geometric representation. Thus the language of synthetic geometry made it possible to grasp only a small fragment of the
realm of algebraic problems, a fragment that was perhaps too narrow
to stimulate the creation of an independent mathematical discipline.

From the cognitive point of view Arabic algebra is not more difficult than Euclid's *Elements*. If we compared the logical structure of
the Arabic algebraic texts with the complex patterns of argumentation
used in Euclid, we would find that it is often much simpler. The reason why the Greeks did not discover algebra cannot be the insufficient
subtlety of their thought. It seems that there must have been some ob-
stacle that prevented the Greeks from entering the sphere of algebraic
thought. The first who entered this new land were the Arabs. There is
no doubt that they learned from the Greeks what is a proof, what is a
definition, what is an axiom. But the Arabic culture was very different from the Greek one. Its center was Islam, a religion which denied
that transcendence could be grounded in the metaphor of sight. There-

fore the close connection between knowledge and sight which formed the core of the Greek *epistéme*, was lost. The Greek word *theoria* is derived from *theoros*, which was a delegate of the polis who had to oversee a religious ceremony without taking part in it. A *theoria* was what the *theoros* saw. This means that for the Greeks a theory is what we see when we observe the course of events without participating in them. Thus the metaphor behind the Greek notion of theoretical knowledge was *view from a distance*. According to this metaphor, in order to get insight into a (mathematical) problem one has to separate oneself from everything that bounds one to the problem and could disturb the impartiality of his view.

This shows that algebra with its symbolic manipulations was alien to Greek thought. Algebraic manipulations are based not on insight but rather on feel. The goal is not to envision the solution (as it was in geometry) but rather to get a feel for how to obtain it, to get a feel for all the different modifications, transformations and tricks, to get a feel for the possibilities offered by the symbolic language. But these possibilities are not actualized; they are not opened to the gaze. In an algebraic derivation at each step we have only one expression at our disposal. We remember different calculations that we performed in the past and we perceive the actual expression against their background. We feel the analogies and similarities; we perceive the different hints that lead us through the jungle of transformations, through the inexhaustible amounts of possible substitutions straight to the result. Algebraic thought is always in flux; its transformations are incessantly proceeding. When Arabic algebra reached Europe, a dialogue started. It was a dialogue between the western spirit and a fundamentally different but equally deep spirit of algebra. This dialogue is led by the effort to visualize, to see; the effort to bring the tricks and manipulations before ones eyes and so to attain insight. As a tool of this visualization a new language has born: the symbolic language of algebra.

The language of algebra was developed gradually by Italian and German mathematicians during the fifteenth and sixteenth centuries (see Boyer and Merzbach 1989, pp. 312–316). The main invention of these mathematicians was the idea of a variable, which they called *cosa*, from the Italian word meaning "thing". They called algebra *regula della cosa*, i.e., the rule of the thing, and understood it as a symbolic language, in which they manipulated letters just as we manipulate things. For instance, if we add to a thing an equally great thing, we obtain two things, what they wrote as $2r$ (to indicate the thing they usually used the first letter of the Latin word *res*). This new language is a return from geometrical construction to symbolic manipulations, from the iconic to the symbolic language.

1.1.3.1. Logical Power – Ability to Prove Modal Predicates

In comparison with the language of elementary arithmetic, the language of algebra has a fundamental innovation – it contains a symbol for the variable. Thus we can say that the algebraists succeeded in transferring the basic advantage of the language of geometry, its ability to *prove general propositions*, into the symbolic language. For instance, it is possible to prove that the sum of two even numbers is even, by a simple formal manipulation with symbols: $2l + 2k = 2(l + k)$. Thus the symbolic language reached the generality of the iconic language of synthetic geometry. This generality was achieved in geometry by using line segments of indefinite lengths, in algebra with the help of variables. Nevertheless, the new symbolic language of algebra surpassed in logical power the geometrical language. For example, let us take the formula for the solution of the quadratic equation

$$x_{1,2} = \frac{-b \pm \sqrt{b^2 - 4ac}}{2a}. \tag{1.1}$$

The parameters a, b, c confer to this formula a generality analogous to that which the line segment of indefinite length bestows to geometrical proofs. But the x on the left-hand side means that the formula as a whole expresses an individual and the components of the formula represent the different steps of its calculation. In this way *the procedure of calculation gets explicitly expressed in language.*

The superiority of the language of synthetic geometry over that of elementary arithmetic is connected with the fact that the particular steps of a geometrical construction are not lost (as the steps of a calculation are lost). Each line or point used in the process of construction remains a constituent of the resulting picture. Nevertheless, what gets lost in the process of construction is the order of its particular steps. This is the reason why geometrical construction is usually supplemented by a commentary written in ordinary language, which indicates the precise order of its steps. It is important to notice that these commentaries do not belong to the iconic language of geometry.

The language of algebra, on the other hand, is able to *express the order of the steps of a calculation within the language.* Thus we need no further commentary on the above formula similar to the one we need on a geometrical construction. The formula represents the process of calculation.[7] It tells us that first we have to take the square of *b*, subtract from it four times the product of *a* and *c*, etc. So the process of solution becomes expressed in the language. The structure of an algebraic formula indicates the relative order of all steps necessary for the calculation. Thanks to this feature of the language of algebra, *modal predicates*, for instance, insolubility can be expressed within the language. The insolubility of the general equation of the fifth degree was proven at the beginning of the nineteenth century by Paolo Ruffini, Niels Henrik Abel and Evariste Galois. It is also possible to prove the insolubility of the problems of trisecting an angle, duplicating a cube, or constructing a regular heptagon. Thus the language of algebra is superior to the language of synthetic geometry. The proofs of insolubility are an illustration of the logical power of the language of algebra.

The language of geometry has no means to express or to prove that some problem is insoluble. The process of solution is something that the iconic language of geometry cannot express. As we have shown in the discussion of the explanatory power of the language of geometry, geometry is able to express the fact that a problem has no solution. Nevertheless, the problem of insolubility is a more delicate one. There is no doubt that for each angle there is an angle that is just one third of it, or that each equation of the fifth degree has five roots. The insolubility does not mean the non-existence of the objects solving the particular problem. It means that these objects, even if they exist, cannot be obtained using some standard methods. The language of algebra is the first language that is able to prove insolubility of a particular problem.

1.1.3.2.   *Expressive Power – Ability to Form Powers of any Degree*

In geometry the unknown quantity is expressed as a line segment of indefinite length, the second power of unknown quantity is expressed as a square constructed over this line segment, and the third power of the unknown is a cube. Three-dimensional space does not let us go further

---

[7]   It is important to realize that algebra is able to express a procedure due to the circumstance that it contains an implicit notion of a function (or as Frege puts it in the passage that we quote on page 13, the mathematicians *"had got to the point of dealing with individual functions; but were not yet using the word, in its mathematical sense, and had not yet formed the conception of what it now stands for"*). It is the distinction between the function and the argument that makes it possible to determine from a formula the exact order in which the operations follow each other.

in this construction to form the fourth or fifth power of the unknown. We characterized this feature of the language of geometry as its expressive boundaries. The language of algebra is able to transcend these boundaries and form the fourth or fifth power of the unknown. We can find traces of geometrical analogies in the algebraic terminology of the fifteenth and sixteenth centuries; for instance, the third power of the unknown is called *cubus*. But nothing hindered algebraists from going beyond this third degree, beyond which Euclid was not allowed by the geometrical space. They called the second degree of the unknown *zensus* and denoted it z. That is why they wrote the fourth degree as zz *(zensus de zensu)*, the fifth as rzz, the sixth as zzz and so on. In this way the symbolic language of algebra transcended the boundaries placed on the language of geometry by the nature of space. Of course, we are not able to say what the fifth power of the unknown means, but this is not important. The language of algebra offers us rules for manipulation of such expressions independently of any interpretation.

The turn from geometrical construction to symbolic manipulation made it possible to discover the method of solving cubic equations.[8] This was the first achievement of western mathematics that surpassed the ancient heritage. It was published in 1545 by the Italian mathematician Girolamo Cardano in his *Ars Magna Sive de Regulis Algebraicis*. The history of this discovery is rather dramatic (see van der Waerden 1985, pp. 52–59). Details about this rather deep result will be presented in Chapter 2.2.2 of the present book. The language of algebra makes it possible to solve problems which in the language of synthetic geometry it is difficult even to formulate.

1.1.3.3.   Explanatory Power – Ability to Explain the Insolubility of the Trisection of an Angle

The language of algebra makes it possible to understand why some geometrical problems, such as the trisection of an angle, the doubling of a cube and the construction of a regular heptagon, are insoluble with

---

[8]   We present here the formula for the solution of the equation $x^3 = bx + c$ in modern notation:

$$x = \sqrt[3]{\frac{c}{2} + \sqrt{\left(\frac{c}{2}\right)^2 - \left(\frac{b}{3}\right)^3}} + \sqrt[3]{\frac{c}{2} - \sqrt{\left(\frac{c}{2}\right)^2 - \left(\frac{b}{3}\right)^3}}$$

Of course Cardano never wrote such a formula. He formulated his rule verbally. Nevertheless, this result illustrates the expressive power of the language of algebra.

ruler and compasses. All problems solvable by means of ruler and compasses can be characterized as problems in which only line segments of lengths belonging to some finite succession of quadratic extensions of the field of rational numbers occur. Thus in order to show the insolubility of the three mentioned problems it is sufficient to show that their solution requires line segments whose length does not belong to any finite sequence of quadratic extensions of the field of rational numbers. This can be easily done (see Courant and Robbins 1941, pp. 134–139). The language of geometry does not make it possible to understand why the three mentioned problems are insoluble. From the algebraic point of view it is clear. The ruler-and-compasses constructions take place in fields that are too simple. This explanation illustrates the explanatory power of the language of algebra.

1.1.3.4.   Integrative Power – Ability to Create Universal Analytic Methods

Euclidean geometry is a collection of disconnected construction tricks. Each problem is solved in a particular way. Thus Greek geometry is also based on memorizing. Instead of memorizing the complete recipes as the Egyptians did, only the fundamental ideas and tricks are to be remembered. However, there are still many of them. Algebra replaces these tricks by universal methods. This innovation was introduced by François Viète in his *In Artem Analyticam Isagoge* (Viète 1591). Before him, mathematicians used different letters for different powers of the unknown $(r, z, c, zz, rzz, \ldots)$, and so they could write equations having only one unknown, whose different powers were indicated by all these letters. Vietes idea was not to indicate the different powers of the same quantity with different letters, but to use the same letter and to indicate its power by a word. He used *A latus*, *A quadratum*, and *A cubus* for the first three powers of the unknown quantity A. Similarly he used *B longitudo*, *B planum* and *B solidum* for the powers of the parameter. In this way the letters expressed the identity of the quantity, while the words indicated the particular power. Thus Viète introduced the distinction between a parameter and an unknown.

Algebraists before Viète worked only with equations having numerical coefficients. This was a consequence of the use of different letters for powers of the unknown. The algebraists were fully aware that their methods were universal, fully independent of the particular values of the coefficients. Nevertheless, they were not able to express this universality in the language itself. Viète liberated algebra from the necessity to calculate with numerical coefficients only. His idea was to express the coefficients of an equation with letters as well. In order not to confuse coefficients with unknowns, he used vowels (A, E, I, O, U) to express unknowns and consonants $(B, C, D, \ldots)$ to express coefficients. So the equation which we would write as $ax^3 - by^2 = c$ he would write as

> *B latus in A solidum – C quadratum in E planum equatur*
> *D quadrato-quadratum.*

It is important to note that for Viète the coefficients had dimensions (indicated by the words *latus*, *quadratum*, or *cubus*), just like the unknowns (indicated by the words *longitudo*, *planum*, or *solidum*), so that all terms of an equation had to be of the same (in our case of the fourth) dimension. So his symbolism was a cumbersome one, and many simplifications were needed until it reached its form used at present. But the basic gain, the existence of universal analytic methods, is already present.

Viètes *analytic art*, as he called his method, was based on expressing the unknown quantities and the parameters of a problem by letters. In this way the relations among these quantities could be expressed in the form of an equation containing letters for unknown quantities as well as for parameters. Solving such an equation we obtain a general result, expressing the solution of all problems of the same form. In this way generality becomes a constituent of the language. The existence of universal methods for the solution of whole classes of problems is the fundamental advantage of the language of algebra. The language of synthetic geometry does not know any universal methods. Geometry can express universal facts (facts which are true for a whole class of objects), but it operates with these facts using very particular methods. Algebra developed universal analytic methods, which played a decisive role in the further development of the whole mathematics. From algebra the analytic methods passed to *analytic geometry* (Descartes 1637) and *mathematical analysis of infinitesimals* (i.e., calculus, Euler 1748), then to physics in the form of *analytic mechanics* (Lagrange 1788) and *analytic theory of heat* (Fourier 1822) till they reached logic in *mathematical analysis of logic* (Boole 1847). Thus the integrative power of the language of algebra permeates vast regions of western thought.

1.1.3.5.   Logical Boundaries – Casus Irreducibilis

Studying equations of the third degree, Cardano discovered a strange thing. If we take the equation $x^3 = 7x + 6$, and use the standard recipe for its solution, we obtain a negative number under the sign of the square root.[9] Cardano called this case *casus irreducibilis*, the insoluble case. In many respects it resembles the discovery of incommensurability. In both cases we are confronted with a situation in which the language fails. The attempts to express this situation in the language led to paradoxes. And in both cases the therapy consists in extending the realm of objects with the help of which the language operates. In the case of incommensurability it was necessary to introduce the irrational numbers, in the *casus irreducibilis* the complex numbers. After the introduction of irrational numbers, incommensurability is no longer paradoxical, it just indicates the fact that the diagonal of the unit square has an irrational length. Similarly, after the introduction of complex numbers the *casus irreducibilis* is no longer paradoxical, it just indicates that the formula expresses the roots of the equation in the form of a sum of two conjugate complex numbers.

Nevertheless, for Cardano the *casus irreducibilis* was a mysterious phenomenon. It took almost two centuries until the square roots of negative numbers were sufficiently understood. It is important to realize that Cardano did not look for the complex numbers. The complex numbers rather imposed themselves on him. Cardano would have been much happier if no *casus irreducibilis* had appeared. The discovery of complex numbers is thus an illustration of the "law" formulated by Michael Crowe:

> "New mathematical concepts frequently come forth not at the bidding, but against the efforts, at times strenuous efforts, of the mathematicians who create them." (Crowe 1975, p. 16)

---

[9]   The equation $x^3 = 7x + 6$ has the solution $x = 3$ (the further solutions $x = -1$ and $x = -2$ were not considered, as the *cosa* cannot be less than nothing). Nevertheless, if we substitute $b = 7$ and $c = 6$ into the formula presented in note 8 on p. 33, we obtain a negative number under the sign of the square root:

$$x = \sqrt[3]{3 + \sqrt{-\frac{100}{27}}} + \sqrt[3]{3 - \sqrt{-\frac{100}{27}}}.$$

1.1.3.6. *Expressive Boundaries – Transcendent Numbers*

Even if algebraists were able to explain why nobody succeeded in solving the problem of the trisection of an angle, the problem of quadrature of a circle resisted algebraic methods. Gradually a suspicion arose that this problem is insoluble as well. Nevertheless, its insolubility is not for algebraic reasons. This suspicion found an exact expression in the distinction between algebraic and transcendental numbers. Transcendental numbers are numbers that cannot be characterized using the language of algebra. The first example of a transcendental number was given by Joseph Liouville in 1851. It is the number:

$$l = \sum_{n=1}^{\infty} 10^{-n!} = 10^{-1!} + 10^{-2!} + 10^{-3!} + 10^{-4!} + 10^{-5!} + 10^{-6!} + 10^{-7!} + \ldots$$

$$= 10^{-1} + 10^{-2} + 10^{-6} + 10^{-24} + 10^{-120} + 10^{-720} + \ldots$$

$$= 0,1100010 \ldots (17 zeros) \ldots 010 \ldots (96 zeros) \ldots 010 \ldots (600 zeros) \ldots 010 \ldots$$

In the decimal expansion of this number the digit 1 is in the $n!^{th}$ places.[10] All other digits are zeros. This means that the digit one is on the first, second, sixth, twenty-fourth, ... decimal places. Even though this number is relatively easy to define, it does not satisfy any algebraic equation. This means that it is a transcendental number – it transcends the expressive power of the language of algebra (see Courant and Robbins 1941, p. 104–107). Liouville's number l illustrates the expressive boundaries of the language of algebra. In 1873 Charles Hermite proved the transcendental nature of e (the basis of natural logarithms) and in 1882 Ferdinand Lindemann proved the transcendental nature of π. Thus l, e, and π are quantities about which we can say nothing using the language of algebra.

1.1.4. Analytic Geometry

Analytic geometry originated from the union of several ideas or even traditions of thought that existed independently of each other for many centuries. The first of them was the idea of *co-ordinates*. In geography co-ordinates have been used since antiquity. One of the highlights of ancient geography was the *Introduction to geography* (*Geógrafiké*

---

[10] The symbol $n!$ represents the product of the natural numbers from 1 to n (thus $4! = 1.2.3.4 = 24$, while $6! = 1.2.3.4.5.6 = 720$).

Hyfégésis) written by Ptolemy around 150 AD. It contained the longitudes and latitudes of more than 8 000 geographical locations, many of which were in India or China. At the beginning of the fifteenth century we can witness a revival of cartography due to the expansion of sea trade. At that time Giacomo d'Angli discovered in a Byzantine bookshop a copy of the Greek manuscript of Ptolemy's *Geography*, which he translated into Latin. It was published in 1477 at Bologna together with the charts drawn by Italian cartographers. Ptolemy's method based on the use of the co-ordinate system has thus spread through all of Europe. From a geometrical point of view it is fascinating to realize that the cartographers were able to draw faithful outlines of whole continents on the basis of the data collected from sailors. Today, of course, everybody knows the form of Africa but only few realize that before space flights nobody could really *see* these outlines. Thus shapes like those of Italy or Africa, obtained by cartographers, are a very special kind of shape. These shapes were very different from ordinary geometrical forms because nobody could see them prior to their construction by plotting the positions of several hundreds of points of the coastline on the chart using a co-ordinate system. Just as in cartography, so also in analytic geometry we create shapes that were unseen before. Instead of sailors, nevertheless, we seek the help of algebra.

So we come to a second tradition that played an important part in the creation of analytic geometry – *algebra*. As we mentioned in the previous chapter, algebra succeeded in overcoming the boundaries of three-dimensional space. When al-Khwárizmí introduced the terms for the powers of the unknown (his *shai*, *mal* and *kab*) he used geometrical analogies. The word *kab* means in Arabic a cube. But in contrast to the ancient Greeks, who stopped after the third power, the Arabic mathematician went further and introduced higher powers. Thus for instance *mal-mal*, *kab-mal* and *mal-mal-mal* stood for the fourth, fifth and sixth power. In contrast to the first three powers, for which we have a geometrical interpretation in the form of a line segment, of a square and of a cube, the fourth, sixth and all higher powers of the unknown lack any intuitive geometrical meaning. But this did not prevent al-Khwárizmí from introducing algebraic operations for these quantities. The works of al-Khwárizmí were in the twelfth century translated into Latin (in 1126 by Abelard of Bath and in 1145 by Robert of Chester). These translations initiated the development of algebra which led to the creation of algebraic symbolism. From the point of view of analytic geometry the creation of polynomial forms (i.e., expressions such as

$x^5 + 24x^3 - 4x + 2$) was of particular interest. It was these forms which "replaced the data of the sailors" in the creation of the shapes of analytic geometry.

The third idea that entered the creation of analytic geometry made it possible to unite the two previous ones. It consisted in a new interpretation of algebraic operations. This idea was introduced by René Descartes. Ever since Euclid the product $x \cdot y$ of two magnitudes (which were represented by line segments) was interpreted as an oblong with the sides x and y, that is, as a magnitude of a different kind than x and y. It is interesting to notice that in his *Regulae ad directionem ingenii* (written between 1619 and 1628) Descartes still interpreted the product of two line segments as the area of the oblong formed by them (see Regula XVIII). Nevertheless, in the definition of the product of three segments $a \cdot b \cdot c$ he wrote that it is better to take the product $a \cdot b$ in the form a line segment of the length a times b. Thus the idea of interpreting the product of two lines not as an area but again as a line segment was present already in this early work of Descartes, even though its importance was not yet fully grasped. Descartes mentioned it only in passing, only as a trick that made it possible to interpret the product $a \cdot b \cdot c$ as an oblong (with one side $a \cdot b$ and the other c) and not as a prism (with the sides a, b, and c). Thus the product of two line segments leads Descartes to a magnitude of a higher dimension and the ingenious trick is used only to prevent the occurrence of volumes and magnitudes of even higher dimension. When in 1637 Descartes published his *Géométrie* as a supplement to the *Discours de la Méthode*, the importance of the new interpretation of the product was fully recognized. The product of two (and of any higher number of) line segments was understood as a line segment. The product $a \cdot b$ was simply the number that indicated the length of the segment. Thus the product and similarly also the quotient of line segments did not lead to a change of dimensionality. In this way Descartes created for the first time in the history of mathematics a system of quantities that was closed under the four arithmetical operations (addition, subtraction, multiplication, and division) and thus, with slight anachronism we can say that he created the first *algebraic field*. In this way he succeeded in overcoming the barrier of dimensionality also in geometry. The critical step in the creation of analytic geometry was that Descartes interpreted the product of two straight lines a and b as a straight line of the length $a \cdot b$. Thanks to this interpretation it became possible to transfer to geometry the fundamental expressive advantage of algebra, its ability to form and combine magnitudes of any degree.

Analytic geometry, i.e., the iconic language with a new way of generating geometrical objects, was created from a combination of the above three ideas. As a starting point we take a polynomial, that is an *algebraic expression*, for instance $x^5 - 4x^3 + 3x + 2$. From the algebraic point of view it is a purely symbolic object without any geometrical interpretation. It was precisely in order to transcend the boundaries laid on Euclidean geometry by three-dimensional space and to become able to form higher powers of the unknown that the algebraists had to give up the possibility of any visual representation of their formalism. They knew how to calculate with polynomials, but they never associated any geometrical form with them. In the second step of the generation of the new geometrical objects we apply to this purely symbolic algebraic expression Descartes' *geometrical interpretation* of the algebraic operations. Thus we take for instance $x = 1$, and the above polynomial will represent a line segment of the length $y = 1 - 4 + 3 + 2$, that is 2 units. If we take $x = 2$, we obtain the corresponding line segment $y = 32 - 32 + 6 + 2$, that is 8, and so forth. In the third step we plot the pairs of values of x and y that we obtained in the previous step, in a *co-ordinate system*. When we plot enough of them, before our eyes something radically new appears; a new curve which before 1637 nobody could see. In a similar way as the shapes of the continents emerged before the eyes of cartographers, a totally new universe of forms emerged before the eyes of mathematicians.

The polynomial, as introduced by the algebraists, was a purely symbolic object, without any geometrical interpretation. Fortunately this loss of visual representation did not last long. In the seventeenth century analytic geometry was developed. In analytic geometry with every polynomial there is associated a curve. In this way all algebraic concepts such as root, degree, etc. acquire geometric interpretation. For instance the degree of a curve can be geometrically interpreted as the maximal number of its intersections with a straight line. The idea of associating a curve to any algebraic polynomial resembles in many aspects the Pythagorean idea of visualization of numbers, which associated geometric forms to arithmetic properties with the help of figurate numbers. *In a way similar to the Pythagorean visualization of arithmetic, analytic geometry visualizes algebra.* In both cases we have to deal with creation of a new iconic language, which incorporates some features of the particular symbolic language.

1.1.4.1. Logical Power – Proof of the Fundamental Theorem of Algebra

Carl Friedrich Gauss in his doctoral dissertation *Demonstratio nova Theorematis omnem Functionem algebraicam rationalem integram unis Variabilis in Factores reales primi et secundi Gradus resolvi posse* from 1799 proved the fundamental theorem of algebra. In his proof Gauss used the plane as a model of complex numbers. The fundamental theorem of algebra says that for every polynomial $p(x)$ there is a complex number α such that $p(\alpha) = 0$. The zero on the right-hand side of this equation is only for historical reasons; in principle we could put there 7 or any other complex number. The zero is there only because before starting to solve a polynomial equation, the algebraists first transformed it to a form having on the right-hand side a zero. But the fundamental theorem of algebra would hold also if we decided to put any other complex number instead of zero on the right-hand side of the polynomial equation $p(x) = 0$. This indicates that the true meaning of the fundamental theorem of algebra is that a polynomial $p(x)$ defines a surjective transformation $z = p(x)$ of the complex planes x onto the complex plane z. Thus this theorem is rather geometrical (or topological) than purely algebraic.

Gauss gave four different proofs of the fundamental theorem of algebra, the first of them in his doctoral dissertation. An elementary proof of this theorem can be found in (Courant and Robbins 1941, pp. 269–271). The fundamental theorem of algebra is an existential theorem that was a precursor of a whole series of similar existential theorems for differential equations, integral equations and for problems in calculus of variations. What is essential for such proofs is firstly the existence of some analytical formalism (polynomial algebra, differential equations, or calculus of variations) by means of which we define the object, the existence of which we are going to prove. The second component of such existential proofs is some domain (complex numbers, smooth functions, etc.) in the realm of which the existence of the particular object is asserted. This domain has usually some topological property of completeness, closeness, or compactness.

It seems that analytic geometry was the first language that contained both these ingredients necessary for a successful existential proof. On one hand it contained the algebraic formalism in the framework of which we can formulate a polynomial equation. On the other hand it contained the complex plane, i.e., a geometrical continuum which was connected with algebraic formalism in a standard way. All previous existential proofs must have happened inside the formalism itself. Basically they had to be constructive; they had to produce a formal expression which represents the object, the existence of which was at stake. In contrast to this in an existential poof like that of Gauss, we proceed on two levels, one symbolic and the other geometrical. Of course, after the nineteenth century many geometrical existential proofs were considered not sufficiently stringent. Nevertheless, it seems that the language of analytical geometry, by building a tight connection between the formalism and the geometrical structure of the complex plane, was the first language that made it possible to prove the existence of solutions for such a wide class of problems as the class of all polynomial equations.

1.1.4.2.   *Expressive Power – Ability to Represent Algebraic Curves of Any Degree*

Analytic geometry brought a new method of generating geometrical pictures. The configuration is constructed point by point, using a coordinate system. This is something qualitatively new in comparison with Euclid. Euclid generated the picture (a term of the iconic language of geometry) with ruler and compasses. This means that Euclid has some basic "mechanical" forms, which he locates on paper. In contrast to this, analytic geometry breaks every configuration into points and plots the independent points separately, point by point. A form can be associated with every polynomial in this way. Descartes invented this new method of constructing curves by solving a problem stated by Pappus. He writes:

> "If then we should take successively an infinite number of different values for the line y, we should obtain an infinite number of values for the line x, and therefore an infinity of different points, such as C, by means of which the required curve could be drawn." (Mancosu 1992, p. 89)

In this way, using the language of algebra, a much richer universe of forms is disclosed, unknown to the Greeks. Looking back from the Cartesian point of view on Euclidean geometry, we can say that with a few exceptions (such as the *quadratrix* of Hippias, the *spiral* of Archimedes, the *conchoid* of Nicomedes, and the *cisoid* of Diocles (Heath 1921, pp. 226, 230, 238, and 264)) the whole of Euclidean geometry deals with quadratic curves only (i.e., curves whose equations are of second degree). The universe of the analytic geometry is qualitatively richer; it contains qualitatively more curves, than the Euclidean universe. Almost every important mathematician of the seventeenth century came up with a new curve; let us just mention Descartes' *folium*, Bernoulli's *lemniscate*, Pascal's *shell*, the *cardioid*, the *astroid*, and the *strophoid*. So the expressive power of the new geometrical language is greater. It is sufficient to take a polynomial expression, choose its coefficients in an appropriate way and a new form emerges, a form that Euclid could never possibly see. Nevertheless, it is important to realize that within the analytic universe we can reconstruct a region which will correspond to the Euclidean: the universe of quadratic curves.

In algebra a formula represents the order of the particular steps of a calculation. This corresponds to construction of separate points of analytic curves. For each point we have to calculate the values of its co-ordinates using an algebraic formula. So far we use the formula in the same way as it is used in algebra: as a relation between two particular numbers. Nevertheless, analytic geometry goes further. By plotting the independent points separately, point after point, a new form becomes visible. None of the separate points itself gives rise to the form. Only if they are all together can we see the form. The algebraic formula itself determines only each single point; putting them all together is the new step taken by analytic geometry. If we gradually change one co-ordinate, and for each of its values we calculate the second co-ordinate according to the algebraic formula, we obtain a curve. Thus the curve not only expresses a *relation* between isolated values of the variables x and y (the ability to express this relation constitutes the logical power of the language of algebra) but the curve also discloses the *dependence* between the two co-ordinates. This is not a functional dependence yet (i.e., dependence of a function on its argument). The concept of function was introduced by Leibniz. In analytic geometry we have just dependence between co-ordinates, which means that the dependence is geometric in nature. Nevertheless, this geometric way of representing dependence was an important step towards the concept of function itself.

The geometric visualization of dependence as dependence between variables resembles the Pythagorean visualization of arithmetical properties using figurate numbers. Just as the line segment of indefinite length, which the Pythagoreans used in their proofs, was a precursor of the concept of variable, the dependence between variables, as used in analytic geometry, is a precursor of the concept of a function. The line segment of indefinite length, and the dependence between variables are part of the iconic languages, while the concepts of variables and functions are constituents of the symbolic language. Nevertheless, the role of the geometrical intermediate states in the formation of the concepts of a variable or function is clearly visible.

1.1.4.3.   Explanatory Power – Ability to Explain the Casus Irreducibilis

Analytical geometry explains why algebraic formulas can lead to paradoxes. The idea stems from Newton. If solving an algebraic equation means determining the intersection points of a particular curve with the x-axis, the universal solvability of all equations would automatically mean that each curve would have to intersect this axis. This is clearly nonsense. Therefore there must be a way in which an algebraic equation does not give rise to intersection points. The appearance of negative numbers under the sign of the square root can prevent an algebraic equation from having a solution.

Thus the *casus irreducibilis* is not a failure of algebra or of the algebraist – on the contrary. Since algebraic expressions determine analytical curves, the formulas giving the solutions of algebraic equations must fail, in some cases, to give curves the freedom not to intersect. In this way the failure of the formulas, which might look like a weakness of the algebraic language from a purely algebraic point of view, is no weakness at all. Neither is it an exceptional case. It must be a systematic feature of all algebraic formulas, in order to give analytic geometry the necessary freedom. Thus the language of analytical geometry explains the failure of the language of algebra, which looked rather odd from a purely algebraic point of view.[11] Here again we have to do with an explanation similar to the explanation given by synthetic geometry of the insolubility of some arithmetic problems. In both cases the geometrical language disclosed the richness of possible situations responsible for the failure of a particular symbolic language. Thus these explanations are not examples of the skill of some mathematicians. They rather disclose an epistemological feature of the language itself, namely

---

[11] The situation is not so simple in the case of Cardano's *casus irreducibilis* because the cubic parabola has three real roots. Thus here we have to deal with a more delicate question. The formula expressing the solutions of a cubic equation represents its real roots as the sum of complex quantities. Nevertheless, this does not contradict the basic fact that the representation of a polynomial by a curve makes it possible to understand phenomena, which from the purely algebraic point of view appear rather puzzling.

its explanatory power. Who, when and under what circumstances, discovered the explanatory power of a particular language is a historical question. But the explanatory power itself is an epistemological fact, requiring philosophical rather than historical analysis.

1.1.4.4.   *Integrative Power – Integration of Geometric Methods*

Descartes published his *Géométrie* in 1637 as an appendix to his *Discours de la méthode*. It comprises three books. The first book, *Problems the construction of which requires only straight lines and circles* opens with a bold claim:

> "Any problem in geometry can easily be reduced to such
> a term that a knowledge of the lengths of certain straight
> lines is sufficient for its construction." (Descartes)

Descartes showed that any ruler-and-compasses construction is equivalent to a construction of the root of a second-degree equation. The core of this part of *Géométrie* is a general strategy for solving all geometrical problems. It consists of three steps: *naming*, *equating* and *constructing*. In the first step we assume that the problem is already solved, and give names to all the lines which seem needed to solve the problem. In the second step we ignore the difference between the known and the unknown, find the relations that hold among the lines, and express them in the form of algebraic equations. In the third step we solve the equations and construct their roots. Thus Descartes managed to introduce the universal analytical methods of algebra into geometry. As an illustration of this can be taken the construction of the *decagon*:

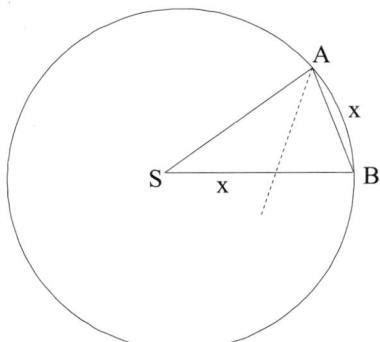

> "Suppose that a regular decagon is inscribed in a circle with
> radius 1, and call its side x. Since x will subtend an an-
> gle 36° at the centre of the circle, the other two angles of

the large triangle will each be 72°, and hence the dotted line which bisects the angle A divides triangle SAB into two isosceles triangles, each with equal sides of length x. The radius of the circle is thus divided into two segments, x and $1-x$. Since SAB is similar to the smaller isosceles triangle, we have $1/x = x/(1-x)$. From this proportion we get the quadratic equation $x^2 + x - 1 = 0$, the solution of which is $x = (\sqrt{5} - 1)/2$." (Courant and Robbins 1941, p. 122)

To construct a line segment of this length is easy. Then we just mark off this length ten times as a chord of the circle and the decagon is produced. We do not need to memorize any construction trick – the constructive part of the problem is trivial. Classical constructive geometry was difficult because to construct any object a specific procedure had to be remembered. In analytic geometry we do not construct objects. Properties of an object are rewritten as algebraic equations, these are solved, and only the line segments with lengths corresponding to the solutions of the equations are constructed. So instead of constructing a regular decagon we have to construct a line segment of the length $(\sqrt{5} - 1)/2$. Thus Descartes brought the universal methods of algebra into geometry.[12]

It is possible that a classical geometer would be able to construct a decagon making fewer steps as we did in the above construction. His construction may seem much more elegant. The advantage of the analytic method is that it is in a sense unrelated to the particular object that we are constructing – in our case the pentagon. With any other object the procedure would be basically the same, only the particular equation would be different. That procedure is based on Descartes' insight that beneath the visible surface of geometry, on which the tricks of the geometers take place, there is a deeper layer of structural relations, which make these tricks possible. The language of algebra is able to grasp this deeper layer of geometry, the structural relations among the mag-

---

[12] Here I have to apologize to the reader. I could not resist the temptation and I have chosen an illustration that is rather elegant. But precisely because of its elegance it obscures the point. In the construction of the regular pentagon I have used a trick – its replacement by the regular decagon. If I were not to do this, I would obtain a similar equation, only not so quickly. But what is important is not how quickly we obtain the result, but the fact that the whole ensuing construction is absolutely trivial. All the difficult work was transferred onto the shoulders of algebra. Algebra led us from the equations to the formula for its solution. For the geometrical part it remains only to construct the roots of the algebraic equations, which is trivial.

nitudes that occur in a particular geometrical construction. The language of analytical geometry discloses a deeper unity which is hidden beneath the apparent diversity of the particular geometrical problems. All problems of classical geometry consisted in the construction of particular line segments, the length of which was determined indirectly by a set of relations to other line segments. The relations determining the particular line segments can be expressed using algebraic equations. Therefore the integrative force of the language of analytic geometry can be characterized by its ability to integrate all the different construction methods of synthetic geometry into one universal scheme.

1.1.4.5.   Logical Boundaries – The Impossibility of Squaring the Circle

The transcendence of π was demonstrated in 1882 by Carl Lindemann. His proof used methods of the theory of functions of complex variables and it can be found in many books (see Stewart 1989, p. 66; Dörrie 1958 p. 148; or Gelfond 1952, p. 54). The transcendence of π means that π is not a root of any algebraic equation. In the problem of the quadrature of the circle the number π plays a crucial role and the transcendence of π means that this problem is insoluble with the methods of analytic geometry. The insolubility of the problem of the quadrature, however, cannot be expressed by means of the language of analytic geometry; it only shows itself in the failure of all attempts to solve it.

1.1.4.6.   Expressive Boundaries – The Inexpressibility of the Transcendent Curves

Soon after the discovery of analytical geometry it turned out that the language of algebra is too narrow to deal with all the phenomena encountered in the world of analytic curves. First of all, two kinds of curves are not algebraic, the exponential curves and the goniometric curves. They cannot be given with the help of a polynomial. These curves transcend the language of classical analytic geometry just as the transcendental numbers transcend the language of algebra.

1.1.5.   The Differential and Integral Calculus

The differential and integral calculus is, just like arithmetic and algebra, a symbolic language, i.e., a formal language designed for manipulation with symbols. It was discovered independently by Gottfried Wilhelm Leibniz and Isaac Newton in the seventeenth century. Nevertheless, the roots of this language reach into ancient Greece, to Archimedes who developed a *mechanical method* for calculating areas and volumes of geometrical figures (Heath 1921, vol. 2, pp. 27–34). This rather ingenious method resembled in many respects our differential and integral calculus. In his calculations Archimedes placed the geometrical object, the area or the volume of which he wanted to determine, on one side of a lever, cut it into thin slices and these slices he then counterbalanced with similar slices of some other geometrical figures. For instance in the calculation of the volume of the sphere he counterbalanced the slices of a cylinder with slices of a sphere and a cone. From the condition of the equilibrium of the lever he then determined the ratio of the areas or volumes of the involved objects. Since in each calculation only one geometrical object had a hitherto unknown area or volume, from the ratio of this unknown area or volume to the known ones Archimedes finally determined the result.

This method is fascinating because in an embryonic form it contains all the ingredients from which the notion of the definite integral will be constructed. First of all, in the method of Archimedes just as in the definite integral, the object, the area or the volume of which we want to determine is cut up into thin slices. Thus in both cases the calculation is understood as a *summation* of slices. This summation happens on an *interval* that Archimedes determined by the projection of the particular geometrical object onto the arm of the lever. The lever itself is also interesting, because it is in a sense the germ of the notion of a *measure*. The different slices of the geometrical object are placed on the arm of the lever not in a haphazard way, but at precise distances (which may be seen to correspond to the elements of dx). Thus with the advantage of hindsight we can say that in Archimedes all the basic ingredients of the notion of definite integral were present.

But equally important as the similarities are also the differences. The first difference is that Archimedes *lacked the notion of a function*. He was not integrating an area delimited by functions, but rather he calculated the volume of objects that were defined geometrically. The language of synthetic geometry is much poorer than the language of functions. Therefore Archimedes calculated only areas and volumes of *single isolated objects*. He was forced to calculate the area or volume of each geometrical object from scratch. For each particular calculation he had to find a special way of cutting it into slices and of counterbalancing these slices with slices of some other geometrical objects. In his calculations he could not use the results of the previous calculations. Therefore the method of Archimedes was a collection of ingenious tricks. In contrast to the geometrical language of Archimedes, the language of the differential and integral calculus is based on the notion of a function. This makes it rich enough for the calculations of different integrals to be tied together by means of particular substitutions and the rule of *per partes*. The result of one integration is automatically the starting point of others. Instead of areas or volumes of isolated geometrical objects we have to do with broad classes of functions, the integrals of which are closely interconnected. We are not forced to start each integration from scratch. The integrals form a systematically constructed calculus that contains entire classes of functions, for which standard methods of integration exist.

The reason for this fundamental difference between the method of Archimedes and the modern integral calculus lies in the language. One could say that Archimedes was forced to invent his tricks to compensate for the weak expressive power of the language of geometry. Only when algebraic symbolism was developed with its formal substitutions, and when on its basis in analytic geometry a much broader realm of geometric forms was made accessible, could Newton and Leibniz introduce the notion of a function and the linguistic framework in which the ideas and insights of Archimedes could be transformed into a systematic technique, which we call differential and integral calculus.

1.1.5.1.   *Logical Power – Ability to Solve the Problem of Quadrature*

Even if analytic geometry brought decisive progress in geometry, it was not able to solve many geometrical problems. Two problems – the problem of tangents, i.e., of finding a tangent to a given curve, and the problem of quadratures, i.e., of finding the area under a given curve – gave birth to technical methods exceeding the language of algebra. These new technical methods, developed in the context of the problem of quadratures, were based on the idea of dividing the given object into infinitely many infinitesimal parts. After an appropriate transformation these parts could be put together again so that a new configuration would be formed, whose area or volume could be determined more easily. Kepler, Cavalieri, and Torricelli were great masters of the new infinitesimal methods. They found the areas or volumes of many geometrical configurations. But for every configuration it was necessary to find a special trick for dividing it into infinitesimal parts and then for summing these parts again. In some cases a regular division led to success, in others the parts had to obey some special rule. Every trick worked only for the particular object, or for some similar ones at best. It lacked a universal language that would enable the discovery of more general techniques.

The basic idea of Leibniz was in many respects similar to the *regula della cosa* of the early algebraists – to create a symbolic language, allowing manipulation with letters (more precisely groups of letters, namely the differentials dx, dy, etc.) mimicking Kepler's, Cavalieri's, and Torricelli's manipulation with infinitesimals. The differential and integral calculus, like algebra or arithmetic, is a symbolic language. It introduces formal rules for manipulation with linear strings of symbols, by means of which it quickly and elegantly gives answers to questions arising in the universe of analytic curves. The central point of differential and integral calculus is the connection, discovered by Newton and Leibniz, between the definite and the indefinite integral:

$$\int_a^b f(x)dx = F(b) - F(a) \qquad \text{where} \qquad F'(x) = f(x). \quad (1.2)$$

This formula makes it possible to *replace the difficult geometrical problem* of quadrature (expressed by the definite integral on the left-hand side, and consisting of the division of the area beneath the curve $f(x)$ into infinitesimally small parts and their rearrangement so that it is possible to determine their sum) by a much *easier computational problem* of formal integration (expressed by the difference of the two values on the right-hand side, consisting of finding for a given function $f(x)$ its primitive function $F(x)$ such that $F'(x) = f(x)$). Actually the formula (1.2) entails that instead of calculating the area given by the definite integral $\int_a^b f(x)dx$, we can first find the primitive function $F(x)$ and then just calculate the difference $F(b) - F(a)$. Therefore if we wish for instance to calculate the area enclosed beneath the curve $y = x^3$ between the boundaries $x = 3$ and $x = 5$, it is not necessary to calculate the integral $\int_3^5 x^3 dx$. It is sufficient to take the function $\frac{1}{4}x^4$, which is the primitive function of x^3 and to calculate the difference $(\frac{5^4}{4} - \frac{3^4}{4}) = \frac{625-81}{4} = 136$. Thus instead of complicated infinitesimal techniques it is sufficient to perform some elementary operations. In most cases finding the primitive function is not so easy, but nevertheless the whole calculation is even then much simpler than geometrical methods developed by Kepler or Cavalieri (see Edwards 1979, pp. 99–109).

The basic epistemological question is: what made possible this fundamental progress in the solution of the problem of quadrature? To answer this question we have to analyze the way which led Newton to the discovery of the formula (1.2). Newton's basic idea was to consider the area below the curve $f(x)$, i.e., the definite integral $\int_a^t f(x)dx$, not as a fixed quantity but as a variable one, i.e., as a function of the upper bound which he identified with the time t.

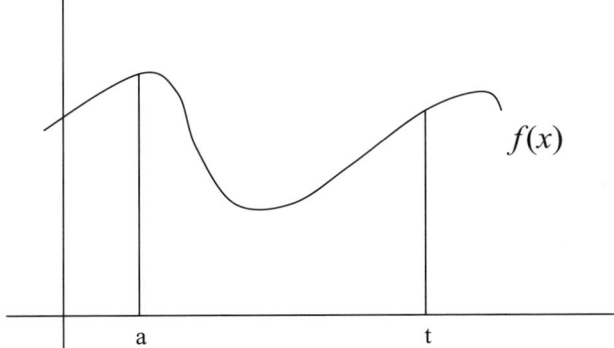

Let us therefore imagine the area beneath a curve $f(x)$, enclosed between $x = a$ and $x = t$, and let the parameter t grow gradually. This means that the right side of the figure will move slowly to the right and thus its area will slowly increase. In order to capture the changing area beneath the curve $f(x)$ Newton introduced the *function* $F(t)$ which expresses this area for any particular moment t:

$$F(t) = \int_a^t f(x)dx .$$ (1.3)

In this way Newton incorporated the problem of the calculation of the area of *one single object* (namely the figure depicted above, but taken with a fixed left side $x = b$) into a whole class of similar problems, one for each value of the variable t. Now instead of trying to determine for a particular value of t the corresponding area, we can analyze the nature of the dependence of the area on the variable t. Among other things we can try to determine the velocity of the growth of the area. Of course, this velocity *depends* on the value of the function $f(x)$ at the particular moment $x = t$. If $f(t)$ is small, then the area will increase slowly, while if $f(t)$ is great, then the area will increase rapidly. We can try to determine this dependence. The velocity of any change is the magnitude of the increment divided by the time in which this increment was achieved. Thus the velocity of the change of area is

$$\frac{\int_t^{t+\Delta t} f(x)dx}{\Delta t} \approx \frac{f(t) \cdot \Delta t}{\Delta t} = f(t). \tag{1.4}$$

That means that the velocity of the growth of the area beneath the curve at a given moment t is equal to the value of the function $f(t)$ at this moment. Thus to calculate the area beneath the curve it is sufficient to find a function such that its velocity of growth is precisely $f(t)$. The velocity is given by differentiation. Thus we can forget about areas. It is sufficient to find a function $F(t)$, whose derivative is $f(t)$. This fact is represented by the formula (1.2).

We see that the decisive step in the discovery of the formula (1.2) was a new view of the area under a curve. Newton looked on this geometrical quantity as a *function*, namely a function of the upper boundary of the figure whose area we calculate. It is important to notice that the concept of the area beneath a curve is already given with the help of a function, namely the function $f(t)$ which determines the particular curve. This is the way in which functions are present in the iconic language of analytic geometry. What is new in the calculus is that the area, which is already determined with the help of the function $f(t)$, is considered to be a function of the upper boundary. Thus we have to deal here with the implicit concept of a function of a function, that is a function of the second degree, as Frege described the calculus in the quoted passage. To be able to deal with the second-order functions (like integrals or derivatives), the ordinary functions had to become explicit. The introduction of an explicit notation for functions is thus a characteristic feature of the symbolic language of the differential and integral calculus, which is fully parallel to the introduction of explicit notation for variables, which occurred in the previous stage of the development of symbolic languages, in algebra.

1.1.5.2. *Expressive Power – Ability to Represent Transcendental Functions*

The differential and integral calculus is a language of higher expressive power than was the language of algebra on which analytic geometry was based. Functions like $\ln(x), \cos(x)$, and elliptic functions are not polynomial. They transcend the boundaries of the language of algebra. For the differential and integral calculus they present no problem. For instance the logarithmic function can be expressed as an infinite series

$$\ln\left(1+x\right) = x - \frac{x^2}{2} + \frac{x^3}{3} - \frac{x^4}{4} + \frac{x^5}{5} - + \ldots , \qquad (1.5)$$

or as an integral

$$\ln(1+x) = \int_0^x \frac{1}{1+t}\,dt , \qquad (1.6)$$

or as a solution of a differential equation

$$y' = \frac{1}{y} , \qquad\qquad\qquad y(1) = 0 . \qquad (1.7)$$

The logarithmic function is a rather simple example. It would be possible to present more complicated examples, for instance Euler's Γ-function, Bessel's functions, Riemann's ζ-function, elliptic functions and a number of special functions occurring in physics or in technical applications. The differential and integral calculus is a symbolic language which can express many functions absolutely inconceivable in the framework of the language of algebra. Nevertheless, the definitions of these new functions use infinite series, derivatives, or integrals, that is, functions of the second degree.

The language of algebra can be embedded in the new language of the calculus. If in an infinite series (expressing for instance $\ln(x)$) we restrict ourselves only to a finite number of initial terms, we obtain a polynomial. In the universe of polynomials, differentiation and integration can be defined by explicit rules, and this universe is closed under these rules. We can consider these restricted operations (prescribing derivative and integral of any polynomial) as new unary algebraic operations.

1.1.5.3.   Explanatory Power – Ability to Explain the Insolubility of the Quadrature of the Circle

Algebra can explain why the trisection of an angle is impossible. Nevertheless, algebra is not able to explain why nobody succeeded in squaring the circle. As we already mentioned, the reason for this is the transcendental nature of the number π. In 1873 Hermite proved the transcendence of the number e and in 1882 Lindemann succeeded, using the ideas of Hermite, in proving the transcendence of π (see Gelfond 1952, p. 54–66). These proofs are based on the language of the differential and integral calculus and so they illustrate its explanatory power. A further interesting fact obtained by these methods is that with the exception of the point $(0, 0)$ at least one of the two co-ordinates $(x, \sin(x))$, that determine a point of the sinusoid, is transcendental. In other words, with the exception of the point $(0, 0)$ none of the points of the sinusoid can be constructed using the methods of analytic geometry. Thus already such a simple curve as the sinusoid totally defies the language of analytic geometry.

1.1.5.4.   *Integrative Power – Mathematical Physics*

The differential and integral calculus is the language that makes it possible for modern physics to unite all its particular branches into an integrated system and so to demonstrate the fundamental unity of nature. While the idea of a universal mathematical description of nature stemmed from Descartes, his technical tools, based on the language of analytic geometry, did not have sufficient integrative power to accomplish his project. The Cartesian polynomials had to be replaced by functions and algebraic equations by differential equations, to integrate all natural phenomena into the universal picture.

Descartes expressed the unity of nature on a metaphysical level. One of the basic purposes of his mechanistic world-view was to unify all natural phenomena. Thus his metaphysics had to fulfil the function which his formal language could not fulfil – to integrate nature into a unified theory. Modern mathematical physics does not require any special metaphysical position in order to see the unity of nature. The unity of physics is fully formal. It is provided by the language and not by metaphysics. Maxwell's equations of an electromagnetic field remained valid after we abandoned the theory of the ether. The ether served for Maxwell as an ontological foundation of his theory, but later it became obvious that the theory could do without this or any other ontological basis. This shows that the unity of physics is a formal unity independent of any ontology. Its source is the integrative power of the language of differential and integral calculus.

1.1.5.5.   Logical Boundaries – Crisis of the Foundations

The first criticism of the foundations of the differential and integral calculus appeared soon after its discovery. In his famous book *The Analyst, or a discourse addressed to an Infidel Mathematician*, which appeared in Dublin in 1734, George Berkeley expressed the view that the whole calculus is based on a series of errors. He criticized the way of reasoning, typical in differential and integral calculus, by which one makes calculations with some quantity assuming that its value is different from zero (in order to be able to divide by it), and then at the end one equates this quantity with zero. Berkeley correctly stressed that if a quantity is zero at the end of some calculation, it must have been zero also at its beginning. Thus all the reasoning is incorrect. According to Berkeley the correct results of calculus are due to compensation of different errors.

Various attempts were presented to rebuild the foundations of the calculus to save it from Berkeley's criticism. Perhaps the most important of them was put forward by Cauchy in his *Cours de l'Analyse* published in 1821. Cauchy tried to base the whole calculus on the concept of limit. Nevertheless, in the rebuilding of the calculus Cauchy left the language of the differential and integral calculus; he left the realm of formal manipulations with symbols and as a fundamental principle by means of which he proved the existence of limits and on which he built the whole theory, he chose the principle of nested intervals. This principle is a geometrical principle. All later attempts to build the foundations of the calculus follow Cauchy in this respect; they leave the language of the differential and integral calculus and erect the building of the calculus on some variant of the theory of the continuum. Thus the crisis of the foundations of the calculus can be seen as a manifestation of the logical boundaries of the language of the differential and integral calculus. An indication of this interpretation is that all attempts to solve this crisis turned their backs on symbolic language.

1.1.5.6.   Expressive Boundaries – Fractals

The successes of the differential and integral calculus justified the belief that all functions can be described with the help of infinite series, integrals, derivatives, and differential equations. Therefore it was a real surprise when the first "monstrous" functions started to appear. Bolzano's function does not have a derivative at any point, Dirichlet's function is discontinuous at each point, and Peano's function fills the unit square. These functions gave rise to a considerable refinement of the basic concepts of differential and integral calculus and gave rise to a whole new branch of mathematics – the theory of functions of a real variable. In the course of their study it turned out that the methods of the calculus can be applied only to a rather narrow class of "decent" functions. The rest of the functions lie beyond the expressive boundaries of the language of the differential and integral calculus.

1.1.6. Iterative Geometry

Differential and integral calculus were born in very close connection to analytical geometry. Perhaps this was one of the reasons why mathematicians for a long time considered Descartes' way of generating curves (i.e., point by point, according to a formula) to be the correct way of visualizing the universe of mathematical analysis. They thought that it would be enough to widen the realm of formulas used in the process of generation, and to accept also infinite series, integrals or perhaps other kinds of analytical expressions instead of polynomials. They believed that the symbolic realm of functions and the iconic realm of curves were in coherence. Leibniz expressed this conviction with the following words:

> "Also if a continuous line be traced, which is now straight, now circular, and now of any other description, it is possible to find a mental equivalent, a formula or an equation common to all the points of this line by virtue of which formula the changes in the direction of the line must occur. There is no instance of a face whose contour does not form part of a geometric line and which can not be traced entire by a certain mathematical motion. But when the formula is very complex, that which conforms to it passes for irregular." (Leibniz 1686, p. 3)

First doubts about the possibility of expressing of an arbitrary curve by an analytical expression occurred in the discussion between Euler and d'Alembert on the vibrating string. The vibrations of a string are described by a differential equation that was derived in 1715 by Taylor. In 1747 d'Alembert found a solution of this equation in the form of travelling waves. Nevertheless, as the differential equation describing the string is an analytical formula, d'Alembert assumed that the initial shape of the string must be given in an explicit form of an analytical expression so that it can be substituted into the equation. Then by solving the equation we obtain the shape of the string in later moments of time. Euler raised against this technical assumption the objection that nature need not care about our analytical expressions. Thus if we gave the string a particular shape using our fingers, the string would start to vibrate independently of whether this initial shape is given by an analytical expression or not. Neither Euler nor d'Alembert was able to bring arguments in favor of his position and thus the problem of the relation between physical curves and analytical formulas stopped soon after it started.

The question of the relation between geometrical curves and analytical formulas got a new stimulus at the beginning of the nineteenth century when Fourier derived the differential equation of heat conduction and developed methods for its solution. Fourier was one of the first scientists who started to use discontinuous functions (in the contemporary sense)[13]. The use of discontinuous functions in mechanics would be absurd; if a function describes the motion of a particle, then its discontinuity would mean that the particle disappeared in one place and appeared in another. Similarly unnatural is the use of a discontinuous function in the theory of the vibrating string. There the discontinuity would mean that the string is broken and thus the vibration would stop. But if a function describes the distribution of heat in a body, then a discontinuous function is something rather natural. It corresponds to a situation of a contact between bodies with different temperatures. So the transition from mechanics to thermodynamics broadened the realm of suitable functions.

Besides a radical extension of the realm of suitable functions, Fourier's *Théorie analytique de la chaleur* (Fourier 1822) contains a method that makes it possible for an almost arbitrary function to find an analytical expression that represents it. Suppose that we have a function $f(x)$ given by means of a chart of its values or of a graph. Fourier's method consists in the calculation of particular numbers (today called *Fourier's coefficients*) by means of (a numerical or graphical) integra-

---

[13] Already Euler used the term "*discontinuous function*" but he used it in a different sense than we do today. For instance, he called the absolute value function, i.e., the function $f(x) = |x|$, discontinuous because on the interval $(-\infty, 0)$ it is given as $f(x) = -x$, while on the interval $(0, +\infty)$ as $f(x) = x$. Euler called a function discontinuous if on different parts of its domain it was defined by means of different analytic expressions.

tion of the function:

$$A_k = \frac{1}{\pi} \int\limits_0^{2\pi} f(x) \cdot \cos(kx) dx \,,$$

$$B_k = \frac{1}{\pi} \int\limits_0^{2\pi} f(x) \cdot \sin(kx) dx \,,$$

$$A_0 = \frac{1}{2\pi} \int\limits_0^{2\pi} f(x) dx \,.$$

Using these coefficients it is possible to express the function $f(x)$ that was formerly given only numerically or graphically, in the form of an analytical expression, today called *Fourier's series:*

$$f(x) = A_0 + \sum_{k=1}^{\infty} A_k \cos(kx) + \sum_{k=1}^{\infty} B_k \sin(kx) \,.$$

In this way Fourier entered the discussion between Euler and d'Alembert on the relation between curves and expressions. Fourier showed that "almost" any function can be represented in the form of an analytical expression. Therefore d'Alembert could answer to Euler, that even if nature does not care about analytical expressions, we can take care of them ourselves. The only problem was the word "almost". A series of prominent mathematicians such as Lejeune–Dirichlet, Riemann, Weierstrass, Lebesgues, and Kolmogorov tried to determine more precisely this "almost". In their works a series of strange functions appeared: Dirichlet's function in 1829, Riemann's function in 1854, Weierstrass' function in 1861, and Kolmogoroff's function in 1923 (Manheim 1964, or Hardy and Rogosinski 1944). The theory of Fourier series enforced the refinement of several notions of mathematical analysis, first of all the notions of function and of integral. In the framework of this theory both Riemann's and Lebesgues' integrals were introduced and finally in the works of mathematicians such as Jordan, Darboux, Peano, Borel, Baire, and Lebesgues a new mathematical discipline – *the theory of functions of a real variable* was born (see Kline 1972, pp. 1040–1051).

As already mentioned, in the study of the functions of a real variable a series of strange objects was discovered. In the nineteenth century these new functions were considered as "pathological" cases (Imre Lakatos will call them "monsters"). Mathematicians still considered the generation of curves point by point in accordance with an analytical formula to be a method which, perhaps with the exception of a few "pathological" cases, gave an adequate geometrical representation of the universe of functions. Towards the end of the nineteenth century these "pathological" functions accumulated in a sufficient number to become the subject of independent study. Mathematicians found their several common attributes, for instance, that a typical "pathological" function has almost nowhere a derivative and has no length. One of the most surprising discoveries was that what originally appeared as pathological exceptional cases were typical for functions of a real variable. This discovery led to a gradual emancipation of the notion of a function from its dependence on analytical expressions. It turned out that the close connection between the notion of a function and its analytical expression led to several distorted views. Nevertheless, as Picard noticed, these distorted views were useful:

> "If Newton and Leibniz had thought that continuous functions do not necessarily have a derivative – and this is the general case – the differential calculus would never have been created." (Picard as quoted in Kline 1972, p. 1040)

Many of the "pathological" functions discovered during the nineteenth century have in common that they are constructed not by Descartes' method of point by point construction in accordance with an analytical formula. They are generated as limits of an infinite iterative process. When in Descartes' method we plot a point on the graph of a function, it never changes. In contrast to this in the new method of generation of geometrical shapes, in each step of iteration the whole curve is drawn anew. The shape is obtained as the limit to which the curves, drawn in the particular steps of the iterative process, converge. Therefore the study of these new forms can be called *iterative geometry*. And just as Descartes' method can be seen as a visualization of the language of algebra (where the central notion is the notion of a polynomial and Descartes ascribed geometrical form to these algebraic objects), iterative geometry can be seen as the *visualization of the language of differential and integral calculus*. The central notion of the calculus is the notion of a limit transition and the new method of generation of geometric forms by means of an iterative process unveils the incredible richness of forms contained in the notion of the limit transition. Thanks to the methods of computer graphics, which make it possible to see the new forms created in iterative geometry on the screen of a computer, the new objects of iterative geometry, often called fractals, became known to a wider public and found their way even into art (see Peitgen and Richter 1986).

1.1.6.1.   Logical Power – the Ability to Prove the Existence of Solutions of Differential Equations

Just as the language of analytic geometry made it possible to prove the fundamental theorem of algebra, according to which every polynomial has at least one root, the language of iterative geometry offers tools for the proof of theorems of existence and uniqueness of solutions for wide classes of differential equations. Differential equations are equations in which unknown functions together with their derivatives occur. It seems that the first differential equation was Newton's second law, which can be written in the form (suggested by Euler) as:

$$F = m\frac{\mathrm{d}^2 x}{\mathrm{d}t^2}.$$

This equation relates the acting force F to the acceleration of the body (i.e., the second derivative of the position x of the body) caused by this force. In the early period of the development of the theory of differential equations, mathematicians wanted to determine the solution of a differential equation in the explicit form of an analytical formula. They developed methods by means of which it was possible to solve wide classes of such equations. In most of these methods they started from the assumption that the solution will be a combination of functions of some special form (polynomial, exponential, etc.) depending on several parameters. Then they substituted these combinations into the *differential* equation and so obtained a system of *algebraic* equations for the parameters. After solving them they could determine the sought solution of the original differential equation.

Relatively early however they discovered that some differential equations also have so-called *singular solutions*. These were solutions which remained undetected by standard methods. A more systematic analysis of the singular solutions was offered in 1776 by Lagrange. Apart from the singular solutions, there were broad classes of (non-linear) differential equations for which the standard methods simply did not work. At the beginning of the nineteenth century these technical problems were complemented by a general skepticism towards symbolic methods. Therefore in his lectures between 1820 and 1830

Cauchy put the emphasis on proving the existence and uniqueness of solutions of differential equations, before he started discussing their properties. Cauchy's methods used in these proofs were improved in 1876 by Lipschitz and in 1893 by Picard and Peano.

These proofs of existence and uniqueness of solutions of differential equations are technically too demanding to be expounded here. For our purposes it is sufficient to draw attention to one aspect of them. In their proofs, Cauchy, Lipschitz, Picard, and Peano used a new technique in constructing the function, which represented the sought solution of the differential equation. Instead of determining the function using some analytical expressions, i.e., the symbolic language of differential and integral calculus (as was common until then), they determined the sought function as a geometrical object generated by means of an iterative process. The *method of successive approximations*, as this new method is called, is akin to the iterative methods by means of which the "pathological" functions were generated. Of course, here the intention of mathematicians was the opposite to that present in the creation of "monsters". Now they restricted the iterative process by different conditions of uniformity of convergence so that the functions obtained as limits of the iterative process were "decent". But this difference is not so important. Whether we use the iterative process to create a "monster" that serves as a counterexample to some theorem, or a "decent" function that is a solution of a differential equation; in both cases we create an object by a fundamentally new method. Instead of plotting points following an analytical expression we employ an iterative process.

When Cauchy presented his proofs, nobody (with the probable exception of Bolzano) realized the radical novelty of his method of constructing functions. The new method served the goal of building a theoretical foundation for the theory of differential equations and so it did not attract suspicion. The revolutionary new technique was used to reach conservative goals – to prove that every sufficiently "decent" differential equation has a unique solution. Iterative geometry attracted attention only later, when it led to new unexpected results. Nevertheless, from the epistemological point of view it does not matter how we use a technical innovation. Its novelty is measured not by the surprise which the new results generate (that is a by-product belonging to the sociology of knowledge) but rather by the epistemological properties of these results. So even if the proof of existence and uniqueness of the solutions of differential equations did not raise much stir, it illustrates the logical power of the language of iterative geometry.

1.1.6.2.   *Expressive Power – the Ability to Describe Fractals*

One of the first mathematicians who realized the problems to which the method of iterative generation of curves could lead was Bolzano. In 1834 he discovered a function that was not differentiable at any point (Sebestik 1992, pp. 417 – 431 and Rusnock 2000, p. 174). This example contradicts our intuition which we developed in the study of curves of analytic geometry. The derivative of a function at a particular point determines the direction in which the curve representing the function sets on when it leaves that point. A function that has no derivative at any point corresponds to a curve that cannot be drawn. If we attach our pencil to a particular point of the curve, we do not know in which direction to pull it. Or more precisely, we cannot pull it in any direction, because the curve changes its course at each of its point and so it does not have even the shortest segment in any fixed direction. The optimism of Leibniz expressed in the quotation at the beginning of this chapter is thus demolished.

The first fractals appeared as isolated counterexamples to some theorems of mathematical analysis. In 1918 Hausdorff found a property which these strange objects have in common. If we calculate their so-called *Hausdorff dimension* we will obtain not an integer, as in case of ordinary geometrical objects, but a fraction or even a real number. The Hausdorff dimension is for a point 0, for an ordinary curve it is 1, for a surface 2, and for a solid 3. The dimensions of fractals are somewhere in-between. So *Cantor's set* has Hausdorff dimension approximately 0.6309; *Koch's curve* 1.2619; and *Sierpinski's triangle* 1.5850 (see Peitgen, Jürgens and Saupe 1992). This is also the origin of the name *fractal* – the name indicates the fractional (i.e., non-integer) value of the Hausdorff dimension of these objects. Later another interesting property was discovered – their selfsimiliarity. The selfsimiliarity of fractals means that when we take a small part of a fractal and magnify it sufficiently, we obtain an object identical with the original one.

Later it turned out that fractals play an important role in chaotic dynamics and in the description of turbulence. So scientists slowly stopped viewing iterative geometry as a vagary of mathematicians and started to see it as an independent language for the description of geometric forms. Thus besides synthetic geometry that generates its objects by ruler and compasses, and analytic geometry that generates its objects point by point according an analytical formula, iterative geometry represents a third kind of language that can be used in the description of geometrical forms. The terms of this language are fractals. To try to describe a fractal by a formula, i.e., to generate it by Descartes' method, is hopeless. It is not difficult to see that the universe of iterative geometry is fundamentally richer than the universe of Cartesian curves. Nevertheless, the universe of analytic geometry can be delineated inside of the universe of iterative geometry as the region that we obtain when we restrict the iterative process by sufficiently strong conditions of uniformity of convergence. The Cartesian world of analytic geometry is the "smooth part" of the universe of iterative geometry – just as the universe of Euclidean curves is the "quadratic part" of the world of analytic geometry.

1.1.6.3. *Explanatory Power – The Ability to Explain the Insolubility of the Three-Body Problem*

In classical mechanics we call the problem of determining the trajectories of the motion of two bodies with masses m_1 and m_2 which interact only by gravitational attraction the *two-body problem*. This problem is also called *Kepler's problem* because Kepler, analyzing the data of the positions of the planet Mars, found (empirically) the basic properties of the solutions of the two-body problem: the elliptical form and the acceleration in the perihelion. Of course in Kepler's days the two-body problem could not even be formulated, because the law of universal gravitation was not known. This law was discovered by Newton who also solved the two-body problem under some simplifying conditions. The complete solution of this problem was found by Johannes Bernoulli in 1710.

The *three-body problem* is analogous to the previous one; the only difference is that instead of the trajectories of two bodies we have to determine the trajectories of three bodies. Nevertheless, all regularities that were present in Kepler's problem (the elliptical form of the trajectories; the simple law describing the acceleration in the perihelion) are totally lost after the transition to three bodies. Despite the efforts of the best mathematicians of the eighteenth and nineteenth centuries, such as Euler, Lagrange, Laplace, and Hamilton, the three-body problem remained unsolved. For all the mentioned mathematicians to solve this problem meant to solve it analytically, i.e., to find explicit formulas that would determine the position of each of the three interacting bodies at every temporal moment. But in a similar way to the quintic equations in algebra, the three-body problem in mechanics defied all efforts of solution. Today we know that this problem is insoluble. But the insolubility of the three-body problem is not caused by the poverty of language, as was the case in algebra. In the case of quintic equations it turned out that the universe of algebraic formulas is simply too poor and does not contain the roots of quintic equations. The insolubility of the three-body problem has a totally different reason – *deterministic chaos*. The discovery of deterministic chaos was one of the most important achievements of the mathematics of the nineteenth century. It was made by Poincaré in November 1889, when he found a mistake in his paper *On the three-body problem and the equations of dynamics*, (Poincaré 1890) for which he won in January 1889 the prestigious Prize of King Oscar II of Sweden. The dramatic circumstances of the discovery, which led to the withdrawal of the whole edition of the issue of *Acta Mathematica* which contained the original version of Poincaré's paper and to a new printing of the whole issue with the corrected version of the paper, at Poincaré's expense, are presented in the literature (see Diacu and Holmes 1996, or Barrow–Green 1997).

For our purposes it is important to realize that the discovery of the so-called *homoclinic trajectory*, which leads to the chaotic behavior of a system of three bodies, was made possible thanks to the language of iterative geometry. Poincaré first introduced a special transformation, the consecutive *iterations* of which represent the global dynamics of the system. He then discovered the chaotic behavior of the system of three bodies thanks to a careful analysis of these iterations. A further area, in which chaotic behavior was discovered, was meteorology. In 1961 Lorenz found chaos in a dynamic system by means of which he modeled the evolution of weather. When the chaotic behavior represented by this model was studied, a remarkable new object, the so-called *Lorenz attractor* was discovered. Lorenz's discovery was followed by several others and so gradually it turned out that many dynamic systems show chaotic behavior – from weather and turbulent flow to the retina of the human eye. In the study of chaotic systems, iterative geometry is used as the basic framework. Therefore we can say that the understanding of chaotic behavior is an illustration of the explanatory power of the language of iterative geometry. More detailed exposition of the theory of chaos can be found for instance in (Peitgen, Jürgens and Saupe 1992).

1.1.6.4.   Integrative Power – The Description of Natural Forms

Fractals were originally discovered as counterexamples of some theorems of mathematical analysis. Therefore their purpose was destructive rather than integrative. Even at the beginning of the twentieth century when fractals had accumulated in a sufficient number so that some of their common properties could be found, they were still more an illustration of the ingenuity and the imaginative force of mathematicians than something useful. When in 1967 in the journal *Science* Mandelbrot's paper *How long is the coast of Britain?* appeared, it turned out that fractals are not a mere creation of the imagination of mathematicians, but can be used in describing natural phenomena. Ten years later in his book *Fractal Geometry of Nature* (Mandelbrot 1977) Mandelbrot drew attention to the fact that many natural objects, such as clouds or trees, in many respects resemble fractals. Thus a series of forms that were formerly ignored by science became subject to mathematical study:

> "Why is geometry often described as 'cold' and 'dry'? One reason lies in its inability to describe the shape of a cloud, a mountain, a coastline, or a tree. Clouds are not spheres, mountains are not cones, coastlines are not circles, and bark is not smooth, nor does lightning travel in a straight line... The existence of these patterns challenges us to study these forms that Euclid leaves aside as being formless, to investigate the morphology of the amorphous. Mathematicians have disdained this challenge." (Mandelbrot 1977, p. 13)

And indeed, when we consider the techniques for generating geometrical forms offered by iterative geometry, we will find that they are able to generate forms that are surprisingly similar to the form of a tree, a leaf of a fern, the line of the seashore, or the relief of a mountain. Thus iterative geometry makes it possible to create faithful representations of natural forms. By means of the language of analytic geometry it would be impossible to achieve anything similar. The close relation of iterative geometry to natural forms is not so surprising, if we realize that every multicellular organism is the result of the iterative process of cell division. Therefore it seems to be natural that the language that generates its objects by iterations of a particular transformation is suitable for description of the forms of living things. The discovery of this (iterative) unity of all living forms can be therefore seen as an illustration of the integrative force of the language of iterative geometry. Where the previous languages saw only unrelated, haphazard, amorphous forms, iterative geometry finds order and unity.

1.1.6.5.   Logical Boundaries – Convergence of Fourier Series

The theory of Fourier series played an important role in the creation of iterative geometry. It supplied a sufficiently rich realm of functions which were constructed as limits of an iterative process. Thus the theory of Fourier series was the birthplace of many examples, notions, and methods of iterative geometry. Nevertheless, it is rather interesting that the question of the convergence of Fourier series enforced the creation of a new language, the language of set theory, because the language of iterative geometry was not sufficiently strong to answer that question. Our experience hitherto with the development of the language of mathematics makes it possible to understand this fact and see it more as a rule than an exception in the evolution of mathematics. The situation with *analytic geometry* was similar. Analytic geometry was created thanks to the Cartesian visualization of the polynomials, but later it turned out that the language of polynomials was too narrow for an adequate description of the phenomena which we encounter in analytical geometry. It was necessary to create the differential and integral calculus which is much more adequate for characterization of different properties of analytic curves. Or we can take the example of the *differential and integral calculus*. This was created in close connection to analytic geometry, but in the end it turned out that if we wish to get an undistorted picture of the fundamental notions of the calculus, it is much better to base it on iterative than on analytic geometry. So it seems that the fragment with the help of which we enter the universe of a particular new symbolic or iconic language (in the case of iterative geometry this fragment was the theory of the Fourier series) is in most cases not appropriate to answer the deep new questions posed by this new universe. The lengths of many simple curves of analytic geometry can be calculated only by means of the differential and integral calculus; similarly the question of existence of a solution of a differential equation can be understood by means of iterations of some (contractive) mapping.

Therefore we should not be surprised that the question of the convergence of Fourier series can only be answered in the framework of set theory. We need first of all the notion of the Lebesgues integral, on which the whole modern theory of Fourier series is based, and this notion presupposes measure theory. Therefore we can say that the question of the convergence of Fourier series transcends the logical boundaries of the language of iterative geometry. The logical power of this language is insufficient to answer that question. Furthermore, between the theory of Fourier series and set theory there is also an interesting historical connection. It was the study of convergence of Fourier series that led Cantor to the discovery of set theory. As was noticed by Zermelo, the editor of Cantor's collected works, in his commentary to the paper *Über die Ausdehnung eines Satzes aus der Theorie der trigonometrischen Reihen*:

> "We see in the theory of Fourier series the birthplace of Cantor's set theory." (Cantor 1872)

1.1.6.6. Expressive Boundaries – Non-Measurable Sets

Even if the richness of the universe of fractals may seduce one to believing that the language of iterative geometry is strong enough to express any subset on the real line, nevertheless, there are sets of real numbers that cannot be expressed by means of this language. These are, for example, the non-measurable sets, the existence of which is based on the axiom of choice. It is precisely the non-constructive character of the axiom of choice that is the reason why the different sets, which can be defined by means of this axiom, cannot be constructed using an iterative process. Thus the existence of non-measurable sets illustrates the expressive boundaries of the language of iterative geometry.

1.1.7. Predicate Calculus

The history of logic started in ancient Greece, where there were two independent logical traditions. One of them has its roots in Plato's Academy and was codified by Aristotle in his *Organon*. From the contemporary point of view the Aristotelian theory can be characterized as a *theory of inclusion of classes* (containing also elements of quantification theory). Thanks to Aristotle's influence during the late Middle Ages, this logical tradition had a dominant influence on the development of logic in early modern Europe. The second tradition was connected with the Stoic school and from the contemporary point of view it can be characterized as *basic propositional calculus* (first of all a theory of logical connectives). In antiquity, probably as a result of the antagonism between the Peripatetic and the Stoic schools, these two logical traditions were considered incompatible. Nevertheless, it is closer to the truth that they complemented each other and together covered a substantial part of elementary logic. The period of intensive development of logic during classical antiquity was followed by a period of relative stagnation, which lasted during the middle ages and early modern period. This stagnation was only partial because, for instance in the field of modal logic, remarkable results were achieved (Kneale and Kneale 1962). Nevertheless, most of the achievements and innovations of medieval logic were almost completely lost as a result of rejection of scholastic philosophy. The authority of Aristotle in the field of logic had as its consequence that even Boole considered the theory of syllogisms to be the last word in logic and saw his contribution only as a rewriting of Aristotelian logic in a new algebraic form.

But even if Boole did not dispute the authority of Aristotelian logic, the importance of his *The Mathematical Analysis of Logic* (Boole 1847) can be seen in bringing into mutual contact classical logic and the symbolic language of algebra. Of course, there were some attempts in this direction earlier (by Leibniz and Euler), but these did not go further than some general proclamations and a few simple examples. It was Boole who created the first functioning logical calculus that is still in use under the name *Boolean algebra*. From the ideas of Boole a whole new logical tradition emerged. This tradition was called the algebra of logic and is represented by names such as de Morgan, Venn, Jevons, and Schröder. In this way, logic that was traditionally considered to be a philosophical discipline, came into close contact with mathematics. This contact turned out to be fruitful not only from the point of view of philosophy but also of mathematics. It took place in a favorable moment of time, which is often called the crisis of the foundations of mathematics. The discovery of the pathological functions (that we discussed in the previous chapter) shattered confidence in the intuitive content of such notions as function, curve, or derivative and led to the conviction that all intuitive reasoning should be removed from the foundations of mathematics. As a replacement for intuitive reasoning, on which until then all of mathematics rested, logic came more and more to the front. But if logic wanted to cope with the new role of securing the foundations of mathematics, it was necessary to change the relation of logic to mathematics.

Logicians like Boole or Schröder considered Aristotelian logic as a correct articulation of the fundamental logical principles. Their aim was only to recast this traditional logic into a new algebraic symbolism and so to make it accessible to mathematical investigations. Thus they wanted to *use the language of mathematics to advance logic.* A radically different view on logic is connected with the names of Frege, Dedekind, and Peano. These mathematicians reversed the above relation when they wanted to *use logic as a tool to advance the foundations of mathematics.* This change of perspective led to a surprising discovery: Aristotelian logic is not only insufficient for construction of the foundations of mathematics (this is not so surprising, because Aristotle lived long before the problems in the foundations of mathematics emerged), but Aristotelian logic is unable to give an analysis of even the most simple propositions of elementary arithmetic, such as

$$2 + 3 = 5.$$

The reason for this is that according to Aristotle every judgment has a subject-predicate structure. It turns out, however, that the above equation does not have an unequivocal decomposition into a subject and a predicate. There are at least 6 ways to do it:

1. The subject is the number 2 and we assert that after adding 3 to it we obtain 5.

2. The subject is the number 3 and we assert that after it is added to 2 we obtain 5.

3. The subject is the number 5 and we assert that it is the sum of 2 and 3.

4. The subject is the sum 2 + 3 and we assert that it is equal to 5.

5. The subject is the addition + and we assert that applied to 2 and 3 it gives 5.

6. The subject is the equality = and we assert that it holds between 2 + 3 and 5.

This ambiguity indicates that the decomposition of a proposition into a subject and a predicate is not an issue of logic but it has rather to do with rhetorical emphasis. Each of our six decompositions lays emphasis on a different aspect of the proposition, which from the logical point of view make no difference. The subject-predicate decomposition of propositions is closely related to the Aristotelian quantification theory. According to Aristotle, the quantification determines the scope within which the predicate is asserted about the subject. Aristotle divides propositions into universal, particular, and singular depending on whether the predicate is asserted about all, some, or a single subject. Therefore in each proposition only one "variable" can be quantified, namely the one that is in the role of the subject. This, of course, cannot be sufficient for the needs of mathematics, where for instance the notion of continuity requires the quantification of three variables.

In 1879 Gottlob Frege published his *Begriffsschrift, eine der arithmetischen nachgebildete Formelsprache des reinen Denkens*, which contained a new formal language, that we call *predicate calculus*. While arithmetic manipulates with numbers, algebra with "letters" and mathematical analysis with differentials, Frege created a calculus for symbolic manipulation with concepts and propositions. Perhaps the most important innovation of Frege in comparison with Aristotelian logic was replacement of the subject-predicate structure of a proposition by the argument-function structure. This made it possible to broaden fundamentally the scope of quantification theory. While Aristotle could quantify only one argument (the subject of the proposition), Frege can quantify as many arguments as the proposition contains. Frege constructed his predicate calculus in an axiomatic way. The system of axioms that he proposed is remarkable, because as was shown by Gödel half a century later, it is complete. And, with the exception of a small blemish that was detected by Lukasiewicz (the axiom 3 can be derived from axioms 1 and 2), Frege's axioms were also independent.

1.1.7.1. Logical Power – the Proof of Completeness of Predicate Calculus

In classical mathematics the notion of a proof was understood intuitively as a convincing and valid argument. Nevertheless, occasionally it happened that a proof that seemed correct to one generation of mathematicians was rejected by the next. This happened for instance with proofs based on manipulations with infinitesimals. Many of Leibniz's or Euler's arguments were some generations later met with suspicion, and mathematicians like Bolzano, Cauchy, and Weierstrass replaced them by totally different argumentation. A similar destiny met even Euclid, who was considered for many centuries a model of logical precision. Pasch noticed that when Euclid constructed the middle point of a line segment, he *used* the fact that the two circles described from the endpoints of the segment with radiuses equal to its length do intersect. Nevertheless, among the postulates and axioms of Euclid there is no one that would guarantee the existence of such an intersection point.[14] The existence of this point is thus a hidden assumption on which, besides its postulates and axioms, Euclidean geometry is erected. Later it turned out that it is by far not the only such assumption.

The manipulation with objects of doubtful status (such as the infinitesimals), or the use of hidden assumptions (such as the assumption of the existence of intersection of circles), are two extremes between which lies a whole range of intermediate positions. Therefore a classical mathematician could be never absolutely sure that the propositions, which he "proved" were really proven. To exclude such doubts was Frege's main objective in the creation of his calculus. In his *Begriffsschrift* he formalized the language of mathematics to such a degree that it was possible to turn logical argumentation into manipulation with symbols. Therefore, in contrast to his predecessors, when Frege proved a proposition he could be sure that nobody would ever discover some mistake in his proof.[15] Frege turned the notion of a proof, which generations of mathematicians used more or less intuitively, into an exact mathematical notion. So we can take as illustration of the logical power of the language of predicate calculus the proof of completeness of the propositional calculus given by Bernays in 1926, and the proof of completeness of the predicate calculus given by Gödel in 1930.

1.1.7.2. Expressive Power – Formalization of the Fundamental Concepts of Classical Mathematics

The expressive power of the language of predicate calculus can be seen in its ability to define the fundamental concepts of classical mathematics. This often required making rather subtle distinctions. So for instance the difference between the pointwise and uniform convergence of a functional series consists only in the order of two quantifiers. Of

---

[14] Pasch's argumentation is based on the achievements of iterative geometry; first of all on Dedekind's paper *Stetigkeit und Irrationale Zahlen*. Dedekind discovered the fundamental difference between the system of all real and the system of all rational numbers with respect to completeness. Pasch realized that a plane from which all points with irrational co-ordinates were be eliminated would still serve as a model of Euclidean geometry (through any two points with rational co-ordinates passes exactly one straight line consisting of rational points, etc.). In that model, nevertheless, the two mentioned circles would not intersect, because the point of their intersection has one irrational co-ordinate and so does not belong to the model. Thus the existence of the intersection of circles cannot be a consequence of Euclid's postulates, because the postulates are satisfied by the model, while the intersection of the circles does not exist there.

[15] It is an irony of fate that when Frege abandoned the system of his *Begriffsschrift* and turned to his *Grundgesetze der Arithmetik*, the new system turned out to be contradictory.

course, mathematicians used these notions several decades before the creation of formal logic. But Cauchy's "proof" of the false theorem that the limit of a convergent series of continuous functions must be continuous shows that even mathematicians of Cauchy's standing were not always able to handle these fine differences correctly (see Edwards 1979, p. 312). Many fundamental notions of the theory of functions of a real variable and functional analysis are rather difficult to master without their, at least basic, formalization. Therefore we are now used to define the basic notions of our theories by means of formal logic. In order to see the advantages of this approach, it is sufficient to look at the history of a sufficiently complex mathematical theory. The development of such theories was often hindered by ambiguity of their fundamental concepts, which were defined in a loose and non-formal way. For this reason A. F. Monna gave his book on the history of the Dirichlet principle the subtitle "*a mathematical comedy of errors*" (Monna 1975).

1.1.7.3. Explanatory Power – Formulation of the Fundamental Questions of Philosophy of Mathematics

In the nineteenth century mathematicians were confronted with different philosophical questions. The problem of Euclid's fifth postulate and the discovery of non-Euclidean geometries raised questions about the relation of mathematical theories to reality, about their truth and necessity. An exposition of this debate can be found in (Russell 1897) and its historical summary in (Torretti 1978). The birth of formal logic made it possible to cast these philosophical questions into a precise mathematical form as the questions of *consistency*, *independence*, and *completeness* of a system of axioms. Similarly in the course of the arithmetization of mathematical analysis the question arose whether arithmetic itself can be further reduced to logic or whether natural numbers form a system of objects that is independent from formal logic, and have to be therefore characterized by a set of extra-logical axioms, just as we characterize the basic objects of elementary geometry. Formal logic made it possible to formulate such questions in an exact mathematical manner and it offered tools for their formal investigation. Our understanding of many problems in philosophy of mathematics was deepened considerably thanks to the methods of formal logic. This advance made in the philosophy of mathematics can thus be seen as an illustration of the explanatory power of the language of predicate calculus.

1.1.7.4. Integrative Power – Programs of the Foundations of Mathematics

The formalization of the language of mathematics is closely connected with three programs of the foundations of mathematics which developed towards the end of the nineteenth century. These programs were originally formulated only for arithmetic as attempts to answer the question of what are numbers. Frege formulated in his *Foundations of Arithmetic* (Frege 1884) the *logicist program*, according to which arithmetic can be reduced to logic. Frege believed that if we formulated the fundamental laws of logic in a sufficiently complete form, it would be possible to derive from them all principles of arithmetic. A few years later Peano in his *Arithmetices principia nova methodo exposita* (Peano 1889) formulated the *formalist program*, according to which numbers are abstract objects independent of logic. For their characterization, just as for the characterization of objects of elementary geometry, it is necessary to make recourse to extra-logical axioms. The third program, initiated by Dedekind in his *Was sind und was sollen die Zahlen?* (Dedekind 1888), can be called the *set-theoretical program*. Dedekind presented a middle position between Frege and Peano. He tried to reduce arithmetic to set theory and therefore defined numbers as cardinalities of sets. Dedekind was in agreement with Peano in characterizing sets as abstract objects by a system of special axioms. On the other hand, he agreed with Frege in that he did not consider numbers to be primitive objects, but reduced them to a more fundamental level, represented by set theory, which he called "*theory of systems*" and considered a part of logic.

Whatever is our opinion on the above programs in the foundations of arithmetic, we cannot deny that all three of them are manifestations of the integrative power of the language of predicate calculus. It was this language which enabled Frege to embark on the project of reduction of arithmetic to logic, and which allowed Peano to formulate his axioms. While working on their programs Frege and Peano made substantial contributions to formal logic. When it later turned out that the three programs cannot be realized in their original version because of the paradoxes discovered by Russell in 1901, all three programs were revised and broadened to include the whole of mathematics. Whitehead and Russell revised and broadened Frege's logicist program in their *Principia Mathematica*. Peano's formalist program was revised and broadened by Hilbert's school at Göttingen – they proposed a formulation of Peano's axioms which was immune to the paradoxes. And

Dedekind's set theoretical program was incorporated by Zermelo into axiomatic set theory. In this way the three programs of the foundations of arithmetic were extended into universal programs including the whole of mathematics. But either in their original version limited to arithmetic or in their advanced and broadened version, the three foundational programs illustrate the integrative power of the language of predicate calculus. The unity that logicism, formalism, and set theory reveal in mathematics is the unity brought to the fore by the means of formal logic.

1.1.7.5. Logical Boundaries – Logical Paradoxes

The principle of nested intervals that Cauchy used as the foundation of mathematical analysis belongs to iterative geometry. It says that under certain conditions an iterative process always has a limit and that this limit is a unique point. Thus in 1821 Cauchy shifted the problem of the foundations of the differential and integral calculus onto the shoulders of iterative geometry. Later, when the "pathological" functions were discovered, mathematicians realized the complexities involved in the notion of an iterative process. Thus even if there were no particular reasons to doubt the principle of nested intervals itself, an effort arose to base mathematical analysis on more solid foundations than this principle can offer. In 1872 Dedekind, Weierstrass, and Cantor independently of each other offered a construction of real numbers and so initiated the arithmetization of mathematical analysis. Their aim was to offer an explicit construction of the real numbers and so to build a foundation on which Cauchy's principle could be proven. These constructions, nevertheless, had one common weakness. They assumed the existence of some actually infinite system of objects (in Dedekind's it was the set of all rational numbers, in Cantor's and Weierstrass' construction it was the set of all sequences of rational numbers). Even though these assumptions seem innocuous, some mathematicians considered the existence of actually infinite systems of objects as not sufficiently clear. Therefore Dedekind (in 1888), Peano (in 1889) and Frege (in 1893), independently of each other, offered three alternative constructions of the system of all natural numbers as a canonical actually infinite set of objects. For a short time it seemed that the project of arithmetization started by Cauchy had reached its definitive and successful end. But soon the logical paradoxes emerged and the foundations of mathematics crumbled.

Russell informed Frege about the paradox in Frege's theory in a letter in 1901 (van Heijenoort 1967 p. 124). Frege was surprised by the paradox, but realized immediately that the same paradox can be formulated also in the system of Dedekind, and it is not difficult to see that the system of Peano has a similar fault (see Gillies 1982, pp. 83–93). This shows that the paradox is not the consequence of some mistake of the particular author. It is not probable that Frege, Dedekind, and Peano would make the same mistake. The conceptual foundations of their systems are so different that the occurrence of the same paradox in all of them can be explained only as a feature of the language itself. The logical paradoxes are not individual mistakes but they rather reveal the logical boundaries of the language. When we characterized the logical power of the language of the differential and integral calculus, we mentioned that the basic logical innovation of this language was the introduction of functions of the second degree. And the paradoxes stem exactly from this source. They are caused by the careless use of second-order functions and predicates. In this respect these paradoxes are analogous to the paradoxes appearing in algebra. In algebra the main logical innovation was the introduction of the (implicit) first order functions (power, square root, etc.) enabling one to express the solution of an algebraic problem in the form of a formula. The paradoxes in algebra (the *casus irreducibilis*) were caused by the careless use of these first-order functions.

The three programs of the foundations of mathematics – the logicist, formalist, and set-theoretical – extricated themselves from the crisis. Nevertheless, this extrication was achieved by means of a stronger language. This resembles algebra, where the extrication from the paradoxes was achieved by means of the stronger language of analytic geometry. This language made it possible to construct a model of complex numbers in the form of the complex plane. Mathematicians used this model to build a semantics for the paradoxical algebraic expressions and to learn to use them safely. In case of the logical paradoxes the situation is similar. Here again a stronger language was invented, the language of axiomatic set theory (or any other equally powerful extensional language), that made it possible to distinguish the paradoxical expressions from the correct ones.

1.1.7.6.   Expressive Boundaries – The Incompleteness of Arithmetic

Despite the fact that large parts of mathematics can be formalized in the framework of logical calculi, it turned out that even this symbolic representation has its limits. These limits were discovered in 1931 by Gödel, who after proving the completeness of the predicate calculus turned to arithmetic and attempted to prove the completeness also of this theory. These attempts resulted in the perhaps most surprising discovery of mathematics in the twentieth century – the discovery of the incompleteness of arithmetic and the improvability of its consistency. Nevertheless, the tools by means of which Gödel achieved his results transcend the language of predicate calculus. Gödel used a new kind of symbolic language, the theory of recursive functions. Therefore the proofs of incompleteness and improvability of consistency do not belong to the framework of predicate calculus. These results can be seen as an illustration of the expressive boundaries of the language of predicate calculus. The situation here is similar to the previous cases. The language of the next stage (in this case the language of the theory of recursive functions, computability, and algorithms) makes it possible to draw the boundaries of the given language. Similarly as the language of algebra made it possible to prove the non-constructability of the regular heptagon, and thus delineated the boundaries of the language of synthetic geometry; or as the language of the differential and integral calculus made it possible to prove the transcendence of π, and so to draw the boundaries of the language of algebra; also Gödel had to use a stronger language than the one, the boundaries of which he succeeded in drawing. In the language itself its boundaries are inexpressible. They only display themselves in the fact that all the attempts to prove, for instance, the consistency of arithmetic undertaken by Hilbert's school were unsuccessful. The language of the predicate calculus, however, did not make it possible to understand the reason for this systematic failure. Only when Gödel developed his remarkable method of coding and laid the foundations of the theory of recursive functions, did he create the linguistic tools necessary for demarcation of the expressive boundaries of the language of predicate calculus.

1.1.8. Set Theory

Infinity fascinated mankind from the earliest times. The distance of the horizon or the depths of the sea filled the human soul with a feeling of awe. When mathematics created a paradigm of exact, precise, and unambiguous knowledge, infinity because of its incomprehensibility and ambiguousness found itself beyond the boundaries of mathematics. The ancient Greeks could not imagine that the infinite (called

$\alpha \pi \varepsilon \iota \rho o \nu$ by them) could become a subject of mathematical inquiry. The Pythagoreans, Plato, as well as Aristotle denied the possibility of a mathematical description of the $\alpha \pi \varepsilon \iota \rho o \nu$. The situation started to change during the middle ages when the attribute of infinity was ascribed to God. This weakened or even eradicated the ambiguity and imperfection traditionally associated with the notion of infinity (see Kvasz 2004). God is perfect and so also must be his attributes, among them infinity. In this way infinity was divested of the negativity that was associated with this notion since antiquity. "The study of infinity acquired a noble purpose; it became a part of theology and not of science" (Vopenka 2000, p. 328). In the Renaissance this new, perfect, and unambiguous notion of infinity started to find its way from theology into mathematics. As an illustration of this process we can mention the *De Docta Ignorantia*, in which Nicholas of Cusa attempted to prove the Trinity using an infinitely large triangle:

> "It is already evident that there can be only one maximum and infinite thing. Moreover, since any two sides of any triangle cannot, if conjoined, be shorter than the third: it is evident that in the case of a triangle whose one side is infinite, the other two sides are not shorter. And because each part of what is infinite is infinite: for any triangle whose one side is infinite, the other sides must also be infinite. And since there cannot be more than one infinite thing, you understand transcendently that an infinite triangle cannot be composed of a plurality of lines, even though it is the greatest and truest triangle, incomposite and most simple.." (Nicholas of Cusa 1440, p. 22)

I quote this text not for analyzing the correctness or the persuasiveness of its arguments. Rather I would like to use it as an illustration of the distance that western thought has travelled since Antiquity. The freedom with which Nicholas of Cusa uses the notion of infinity is amazing. Such a text could not have been written by any philosopher of ancient Greece. After theology had broken the barrier that separated mathematics from the notion of infinity, a gradual transformation of all mathematics started. Euclid's $\varepsilon \iota \vartheta \varepsilon \iota \alpha$ (straight line), which had only a finite length, was replaced by our *straight line*, i.e., by an object of infinite extension. The Greek geometer needed his second postulate ("*To produce a straight line continuously in a straight line*") in order to secure the possibility of extending the $\varepsilon \iota \vartheta \varepsilon \iota \alpha$ as far as he wished;

on the other hand in the Renaissance the straight line "reached the infinity". Similarly the atoms of Democritus witnessed a revival in the form of indivisibles by Kepler and Cavalieri and were finally replaced by infinitesimals in the seventeenth century.

The penetration of the notion of infinity into mathematics was made possible by the change of attitude to this notion that appeared first in theology. Nevertheless, the notion of infinity was from the beginning accompanied by criticism. Many mathematicians and philosophers of the seventeenth and eighteenth century considered manipulations with infinitesimals to be doubtful or even wrong (we can mention Descartes and Berkeley). Therefore relatively early a countermovement started, the aim of which was to eliminate infinitesimals from mathematics. After the first attempts (Carnot 1797, Lagrange 1797) a successful way of eliminating infinitesimals was found by Bolzano in his *Rein analytischer Beweis* (Bolzano 1817) and fully developed by Cauchy in his *Cours d'Analyse de l'École Polytechnique* (Cauchy 1821).[16] But as we already mentioned, Cauchy and Bolzano based their method of elimination of infinitesimals on the intuitive notion of the continuum. Half a century later Dedekind, Cantor, and Weierstrass presented three constructions of the continuum and so brought Cauchy's project to a consummation. These constructions of the continuum, even if independent, all assumed the existence of some infinite system of objects. This indicated that the notion of infinity, which came into mathematics from theology, cannot be so easily eliminated. Even if Dedekind, Cantor, and Weierstrass succeeded in eliminating the infinitesimals, they succeeded at the price of introducing infinite systems of objects as actually existing.

Dedekind analyzed the notion of an infinite system in *Was sind und was sollen die Zahlen* (Dedekind 1888), where he introduced a definition of the infinite set, which we use until now (a set is infinite if it can be mapped onto its proper subset by a one-to-one mapping). Cantor was led to the study of infinite sets by his investigations of Fourier series. His main contribution to set theory was the *Grundlagen einer allgemeiner Mannigfaltigkeitslehre*. He gave here a definition of the notion of a set:

---

[16] When we look into the *Course* we find out that Cauchy preserved the notion of an infinitesimal. Nevertheless, he changed its meaning. An infinitesimal for Cauchy is not an infinitely small number, as it was for his predecessors, but a variable that converges to zero. Thus the foundations of Cauchy's approach were built on the notion of a limit.

"In general, by a manifold or a set I understand every multi-
plicity which can be thought of as one, i.e., every aggregate
of determinate elements which can be united into a whole by
some law. I believe that I am defining something akin to the
Platonic $\varepsilon\iota\delta o\varsigma$ or $\iota\delta\varepsilon\alpha$ as well as to that which Plato called
$\mu\iota\chi\tau o\nu$ in his dialogue Philebus or the Supreme Good."
(Cantor 1883, p. 204; Ewald 1996, p. 916)

Cantor's work on the theory of infinite sets received at the beginning only minimal support and many influential mathematicians,
such as Kummer or Kronecker, opposed it (Dauben 1979, pp. 133–
140). The reason for this opposition lay at least partially in Cantor's
turning against the general trend of mathematics of his times, which
consisted in the elimination of infinity from mathematics. This trend
strengthened after the discovery of Russell's paradox in 1902. Many
mathematicians believed that the notion of infinity (introduced into
mathematics from theology) is alien to the nature of mathematics and
should be eliminated. As an illustration of this attitude we can mention
Poincaré's words:

"There is no actual infinity. The Cantorians forgot this, and
so fell into contradiction. It is true that Cantorianism has
been useful, but that was when it was applied to a real problem, whose terms were clearly defined, and then it was possible to advance without danger.." (Poincaré 1908, p. 499)

Russell's paradox is usually presented as the paradox of the set of
all sets. Therefore an impression could emerge that it is the paradox of
Cantorian set theory. Nevertheless, Cantor was fully aware of the problematic nature of the system of all sets, which he called the *Absolute*.
Thanks to a theological interpretation of the Absolute, which he identified with God (Dauben 1979, pp. 120–148), Cantor avoided the formulation of any mathematical propositions about this notion. Thus the
theological interpretation of his theory saved Cantor from paradoxes.
The other two foundationalist approaches lay open to the full brunt of
the logical paradoxes. Thus Poincaré's attack against the "Cantorians"
was unjustified; the same paradoxes appeared also in the theories of
Frege and Peano.

Despite criticism, set theory slowly started to gain popularity among
mathematicians working in the field of the theory of functions of a real
variable, measure theory, and general topology. These theories witnessed a rapid growth at the end of the nineteenth and the beginning of

the twentieth centuries and so more and more mathematicians started to use the notions and methods of set theory. In 1908 Zermelo in his *Untersuchungen über die Grundlagen der Mengenlehre* found a way to avoid the paradoxes. Analyzing the works of Dedekind and Cantor, Zermelo formulated as axioms the rules that are necessary to form new sets. But he formulated these axioms so that they did not allow the construction of any paradoxical object, analogous to the set of all sets. A discussion of the axioms of set theory can be found in (Fraenkel and Bar-Hillel 1958). Thanks to the work of Zermelo, set theory was consolidated during the short period of six years after the discovery of the paradoxes[17] and became an important mathematical discipline with remarkable results and methods.

The next important shift in set theory appeared in 1914 when Hausdorff's *Grundzüge der Mengenlehre* appeared. This book summarized the results that had been achieved in set theory up to the date of its publication. But Hausdorff's book contains also a fundamental innovation – the notion of a function was there for the first time defined as a *set* of ordered pairs. Dedekind, Cantor, and Zermelo considered functions and sets as two fundamentally different kinds of things. For Hausdorff they were analogous. Hausdorff's definition of an ordered pair was a bit cumbersome. It was based on the assumption of the existence of two special objects, which Hausdorff indicated by the symbols 1 and 2. Using them he defined an ordered pair as $\{\{a, 1\}, \{b, 2\}\}$. The modern definition of an ordered pair as $\{\{a\}, \{a, b\}\}$, i.e., using no special objects, was introduced by Kuratowsky in 1921. Hausdorff's idea of interpreting functions as special sets presented a fundamental step towards a unification of mathematics on the basis of set theory. Set theory became the language in which almost the whole of mathematics is developed.

1.1.8.1. Logical Power – Proof of the Consistency of the Infinitesimal Calculus

Even though the roots of set theory go back to the program of arithmetization of mathematical analysis, which had as its aim to eliminate infinitesimals from mathematical analysis, it is interesting to notice

---

[17] In the framework of the logicist approach Russell proposed a solution of the logical paradoxes based on the theory of types (Russell 1908). Nevertheless, the theory of types was not accepted by the majority of the mathematical community.

that set theory led to a discovery, which made it possible to put the infinitesimals on solid foundations. The first construction of the system of non-standard real numbers was given by Robinson in his paper *Non-standard analysis* (Robinson 1961). Robinson's model was based on the notion of an ultrafilter, the existence of which follows from the axiom of choice. Therefore, construction of the hyperreal numbers as well as proof of the consistency of the theory of infinitesimals, which follows from this construction, can be seen as an illustration of the logical power of the language of set theory. By means of the language of set theory (and of model theory, which is based on this language) it became possible to lay logical foundations beneath many of Leibniz's and Euler's "proofs". These "proofs" were considered as unsound until Robinson showed that they were correct and so we can spare the quotation marks. Nonstandard analysis that grew out from Robinson's construction became in the meanwhile a theory with many important results and applications (see Albeverio et al. 1986, and Arkeryd et al. 1997).

1.1.8.2. *Expressive Power – Transfinite Arithmetic*

One of Cantor's surprising discoveries was his realization that in the successive construction of the so-called *derived sets* of a given set P of real numbers it is possible to continue also after we have made this construction an infinite number of times. At this point it is not important what precisely this operation means (its explanation can be found in Dauben 1979, p. 41). It is important rather that after Cantor created the *first derived set* P', and by the same construction formed from the first derived set the *second derived set* P'', the third derived set P''', etc. he came to the idea of prolonging the steps of construction of the derived sets beyond the limit of any number of steps that can be counted by natural numbers. On the technical level it is not so difficult, because among the derived sets P', P'', P''', ... there is an interesting relation. Each set $P^{(n)}$ is a subset of the previous one. Therefore if we have a finite series of derived sets P', P'', ..., $P^{(n)}$, the last member of this series (which in this case is $P^{(n)}$) can be expressed as the intersection of all of the members of the series:

$$P^{(n)} = \bigcap_{k=1}^{n} P^{(k)} . \tag{1.8}$$

When instead of a finite sequence of derived sets P', P'', ..., $P^{(n)}$ we take an *infinite* one P', P'', P''', ..., $P^{(n)}$, ..., Cantor's idea was to build an analogous intersection of all members of the sequence. Of course, an infinite sequence does not have a last element and therefore the intersection of all of its members will not be equal to some derived set $P^{(n)}$ as it was in the case of the finite sequence. But despite the fact that we do not know from the very beginning what the result of this intersection will be (in contrast to the finite case, where it was sufficient to look at the last member of the series and we knew what the intersection was), the intersection is a well-defined set also in the case of the infinite series. A point x belongs to this intersection if and only if it is a member of each of the sets $P^{(k)}$. Cantor used the symbol $P^{(\infty)}$ to represent the intersection of an infinite sequence of derived sets:

$$P^{(\infty)} = \bigcap_{k=1}^{\infty} P^{(k)} . \tag{1.9}$$

After he had introduced the derived set of an infinite degree, he could start to create the derived sets of this set and thus to create the sets $P^{(\infty+1)}$, $P^{(\infty+2)}$, $P^{(\infty+3)}$, ..., until he reached the next infinite case. Then he could turn to the operation (1.9) and create $P^{(2\infty)}$. In this way the paper *Über unendliche lineare Punktmannigfaltigkeiten 2* (Cantor 1880) brought a decisive shift in the history of mathematics. Cantor made here the first step towards *transfinite arithmetic*. In the paper he used the symbols $\infty, \infty + 1$ or 2∞ to represent the steps of the process of creating derived sets, thus they were *indices*. His attention was directed towards the set P and he wanted to understand what happens with it by the successive derivations.[18] In his later papers Cantor replaced the symbol ∞ by the last letter of the Greek alphabet ω and interpreted this symbol not as an index but as a transfinite *number*. He introduced the distinction between ordinal and cardinal numbers, introduced for both of them the operations of addition and multiplication and created thus the transfinite arithmetic.

---

[18] The fact that Cantor discovered transfinite arithmetic in the study of iterations of a particular operation confirms the connection of set theory to iterative geometry. The iterative processes, by means of which mathematicians of the nineteenth century constructed their strange objects, had steps that could be numbered by natural numbers. Cantor prolonged the iterative process into the transfinite realm. If we examine what enabled Cantor to make this radically new step, we will find out that it was the operation of the intersection of an infinite system of sets and the understanding that an element belongs to this intersection if it belongs into *each one* of the intersected sets. I do not want to indicate that Cantor read Frege's *Begriffsschrift* (it appeared just one year before Cantor's paper). But I would like to stress that the increase of the logical precision in the foundations of the calculus and the parallel development of formal logic during the nineteenth century, which culminated in Frege's work, played a fundamental role also in Cantor's breakthrough into the transfinite realm.

Transfinite arithmetic is doubtlessly one of the most remarkable achievements of set theory. Mathematicians before Cantor were unaware that it is possible to discriminate different degrees of infinity or that it is possible to associate with them (cardinal or ordinal) numbers which can be added and multiplied like ordinary numbers. Therefore transfinite arithmetic is a good illustration of the expressive powers of the language of set theory.

1.1.8.3.   *Explanatory Power – Unveiling the Typicality of the Transcendent Numbers*

As we already mentioned, the first transcendental number was discovered in 1851 by Liouville. In 1873 Hermite showed that also the number e is transcendental, and in 1882 Lindeman proved the transcendence of the number π. Thus the transcendental numbers slowly accumulated. Nevertheless, it still seemed as if the transcendental numbers were some rare exceptions and the overwhelming majority of real numbers were algebraic. Even though Lindeman proved a stronger result than just the transcendence of π, and created an infinite set of transcendental numbers (see Dörrie 1958, or Gelfond 1952), but he constructed the transcendental numbers by means of the algebraic ones, and so still nobody could suspect that there are fundamentally more transcendental numbers than there are algebraic ones.

Therefore when Cantor in 1873 proved that transcendental numbers form an uncountable set, while the algebraic numbers are only countably many, it turned out that exceptional numbers are rather the algebraic ones, and that a typical real number is transcendental. Even though Cantor's proof was a non-constructive one, and so it could not be used to find a new transcendental number, it was a remarkable discovery. After a short time it was followed by similar results, as for instance that a typical continuous function of a real variable has a derivative almost nowhere. Thus under the influence of set theory our view of the real numbers as well as of the functions of a real variable changed in a radical way. The objects which mathematicians of the nineteenth century considered exceptional turned out to be typical; and as exceptional (at least from the point of view of cardinality and measure) must be considered the objects of classical mathematics, such as algebraic numbers, differentiable functions or rectifiable curves. Set theory thus opened a new perspective on the universe of classical mathematics; it changed our view of which mathematical objects are typical and which are exceptional. This change is an illustration of the explanatory power of the language of set theory.

1.1.8.4.   Integrative Power – The Ontological Unity of Modern Mathematics

Even though the first foundationalist program in mathematics was Frege's logicism, while set theory did not have at the beginning such broad ambitions, the truth is that the overwhelming majority of contemporary mathematics is done in the framework of set theory. Therefore while the axiomatic method unifies mathematics on the methodological level, set theory unifies it on the ontological one. If we take some mathematical object – be it a number, a space, a function, or a group – contemporary mathematics studies this object by means of its set-theoretical model. It considers natural numbers as cardinalities of sets, spaces as sets of points, functions as sets of ordered pairs, and groups as sets with a binary operation. Thanks to this viewpoint, mathematics acquired an unprecedented unity. We are used to it and so we consider it as a matter of course, but a look into history reveals the radical novelty of this unity of the whole of mathematics. We are justified in seeing the ontological unity of modern mathematics as an illustration of the integrative power of the language of set theory.

1.1.8.5.   Logical and Expressive Boundaries

Set theory is one of the last re-codings which were created in the history of mathematics and so today a substantial part of all mathematical work is done in its framework. Therefore it is difficult to determine the logical and expressive boundaries of its language. The logical and expressive boundaries of a particular language can be most easily determined by means of a stronger language, which transcends these boundaries and so makes it possible to draw them. This is so, because the stronger language makes it possible to express things that were in the original language inexpressible. Nevertheless, it seems that mathematics has not surpassed the boundaries posed by the language of set theory. Therefore to characterize these boundaries remains an open problem for the future. What we can say today is that from the contemporary point of view the expressive, logical, explanatory, and integrative force of the language of set theory is total. The boundaries of the language of set theory are the boundaries of the world of contemporary mathematics and as such they are inexpressible.

1.2. Philosophical Reflections on Re-Codings

Analysis of the development of the symbolic language of mathematics
from arithmetic and algebra through the differential and integral calculus to the predicate calculus, presented in the previous chapter, can
be seen as an unfolding of the idea from Frege's paper *Funktion und
Begriff*, quoted on page 15. Nevertheless, our exposition differs from
Frege in two respects. First is terminological – we do not subsume algebra or mathematical analysis under "*arithmetic*", but consider them
as independent languages. More important, however, is that we show
how the "*development of arithmetic*", described by Frege, interplayed
with the development of geometry. Frege separated arithmetic from
geometry and connected the different phases in the "*development of
arithmetic*" only loosely, using phrases as "*then they went on*", "*the
next higher level*", or "*the next step forward*". But the question why
'*they went on*' or where "*the next higher level*" came from remained
unanswered. From the point of view of the logicist program this is
perhaps unavoidable, because logicism makes a sharp distinction between the context of discovery and the context of justification. Even
though we do not want to question this distinction, we believe that it
is interesting to try to understand the dynamics of the transitions described by Frege. It turns out that in these transitions a remarkable tie
between the symbolic and the iconic languages appears. The transition
to "*the next higher level*" in development of the symbolic language
happens by means of an iconic intermediate level. Thus for instance
the notion of a variable, the introduction of which represented the transition from arithmetic to algebra, was created in two steps. The first
of them was geometrical and consisted in the creation of the notion of
a segment of indefinite length. The second step occurred when for this
embryonic idea of a variable an adequate symbolic representation was
found. An analogous intermediate geometrical step can be found also
in the creation of the notion of function. This notion too was created
in two steps; the first of them being the birth of an embryonic idea of
function in analytic geometry as the dependence between two variables
represented by means of a curve. The second step occurred when for
this embryonic geometrical notion an adequate symbolic representation
was created. I believe that this interplay between symbolic and iconic
languages deserves some philosophical reflection.

The following table contains an overview of the symbolic and iconic
languages in the development of mathematics, presented in the histori-

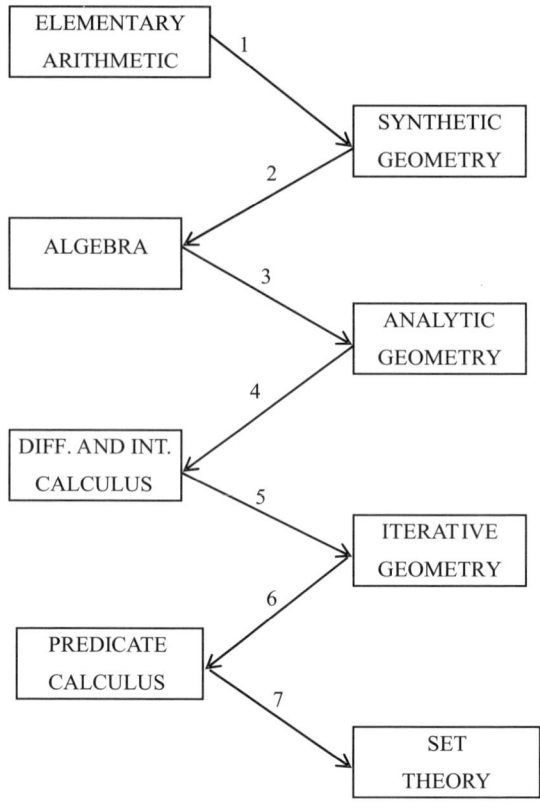

cal order of their appearance. The transitions between these languages consisted in the discovery of a new universe of objects. Four of them, represented by the arrows 1, 3, 5, and 7, correspond to the construction of new iconic languages, each of which opens a door into a new universe of geometric forms. These four transitions consist in the *visualization* of a particular notion that had formerly purely a symbolic meaning. The *Pythagorean visualization* of number by means of figural numbers (represented by the arrow 1) changed in a fundamental way our understanding of the phenomenon of quantity. Numbers that were originally accessible only through counting entered, thanks to the Pythagorean visualization, into a whole range of new relations which led to the creation of a new branch of mathematics called number theory. Number theory is something very different from counting. Instead of an operational relation between the operands and the result of an arithmetical operation (as for instance "*Take away 1/9 of 10, namely 1 1/9; the remainder is 8 2/3 1/6 1/18*", known from the Rhind papyrus) in number theory we have structural relations (as for instance that the sum of two even numbers is even). Similarly in the *Cartesian visualiza-tion* of polynomial forms by means of analytic geometry (represented by the arrow 3), to every polynomial that was until then interpreted purely as a formula and was subjected only to symbolic manipulations, a geometric form was associated. Thanks to this notion, polynomi-als enter into a whole range of new relations, which were until then unimaginable. The solution of a polynomial, which was traditionally one of the main purposes of algebra, obtained a geometrical interpre-tation. Thanks to this interpretation it became comprehensible why certain polynomials have no solution in the domain of real numbers – the corresponding curve simply does not cut the x-axis. In this way the modal predicate of insolubility of a polynomial was transformed into an extensional predicate of the curve (the non-existence of a particular intersection). Similarly, iterative geometry brought the *visualization of the limit transition* (represented by the arrow 5). If we understand the properties of objects such as the Peano curve or the Mandelbrot set, we acquire a better insight into many problems that can occur on the transition to a limit. And it is possible to see also the birth of set theory as a "visualization", or perhaps in this case it would be more appro-priate to speak about the *Cantorian extenzionalization* of the predicate calculus (represented by the arrow 7).[19]

---

[19] The reader may wonder why set theory appears in the right-hand column of the diagram, alongside with synthetic, analytic, and iterative geometry. It may seem more appropriate to place set theory between the two columns because it is not so closely connected to spatial intuition as the other three iconic languages. I have several reasons for this placement. *First of all* there is an analogous relation between set theory and iterative geometry to that between algebra and the differential and integral calculus. Just as an infinite series can be interpreted as a prolongation of a polynomial to "infinity", so also Cantor arrived at his transfinite sets by prolonging the iterative process used in iterative geometry beyond the set of indexes (or steps of iteration) numbered by natural numbers. Therefore set theory is related to iterative geometry in the same way as subsequent languages of the same kind (iconic or symbolic) are related to each other, namely by *using an infinite number of operations of the previous language as one step of the new language.* In an analogous way a curve in analytic geometry is constructed point by point according to a formula. This, from the viewpoint of synthetic geometry, would require an infinite number of construction steps, and therefore it would be impossible to do. In analytic geometry we take this infinite number of construction steps as one step of the new language and take the curve as if it had already been constructed. *The second reason* is that when we define new sets of the form $\{x \in A; \phi(x)\}$, using the axiom of separation, where $\phi(x)$ is a formula of a particular fragment of the predicate calculus, this move is fully analogous to Descartes who defined a curve as a set of points that fulfill a particular polynomial form $p(x)$. In both cases formulas of a symbolic language (predicate calculus, algebra) are used to *define a new object as the system of elements that satisfy the particular formula.* Thus also the relation of set theory to the previous symbolic language follows a rather standard pattern. *And finally* the other iconic languages are also not so directly connected to intuition as it may seem. To believe that one is able really to see a

The three remaining transitions represented by the arrows 2, 4, and 6 correspond to the creation of new symbolic languages, each of which opens a door into a new universe of formulae. These transitions can be characterized as symbolizations in which a particular aspect of the iconic language acquires a symbolic representation. Thus in the birth of algebra (represented by the arrow 2) the *symbolization of the notion of a variable* occurred. An embryonic idea of a variable was already present in synthetic geometry in the form of a line segment of indefinite length. Euclid used this idea in his proofs in number theory. The idea of a line segment of indefinite length is not a fully fledged notion of a variable, because it is closely related to geometrical constructions. The operations which can be performed by such line segments are limited by the three dimensions of space (thus only first, second, and third powers of it can be formed) and by the principle of homogeneity (thus different powers cannot be added or subtracted). Only when the line segments were replaced by letters did the generality of reference which is the core of the notion of a variable obtain its full strength. Similarly the creation of the differential and integral calculus (represented by the arrow 4) brought a *symbolization of the notion of a function*. In an embryonic form the idea of a function is present already in analytic geometry in the form of a curve associated with a polynomial. But this idea lacked an adequate symbolic representation and thus the range of operations which it was possible to do with these curves was rather restricted. Only after Leibniz created a new symbolic language which contained a universal symbolic representation of the notion of a function did it become possible to perform symbolic operations with functions, as for instance to form a function of a function, to perform integration by the method of *per partes*, or integration by the method of substitution. Both of these integration methods were, of course, unimaginable when functions were represented only geometrically. As the last symbolization I would like to mention the *symbolization of quantification and of logical derivation* (represented by the arrow 6) in the predicate calculus. This step followed after a radical increase of precision of the mathematical language that was enforced by problems in analysis. The $\varepsilon - \delta$ analysis of Weierstrass contained in an implicit form many aspects of the quantification theory, and its fine

---

fractal is naïve. And analytic geometry studies also curves in many-dimensional complex or projective spaces. Thus set theory is surely not the first iconic language which does not have a straightforward visualization.

distinctions between different kinds of convergence gave rise to many notions, which found their explicit expression in the predicate calculus.

The changes of languages that have the form of *visualization* or *symbolization* are closely related and I would like to subsume them under the common notion of *re-codings*. Our analysis of re-codings revealed a remarkable regularity of alternation of the symbolic and the iconic languages, which is important, both for the philosophy of mathematics and for mathematics education.

1.2.1. Relation between Logical and Historical Reconstructions of Mathematical Theories

Understanding of the role of the iconic intermediate levels in the development of mathematical symbolism could shed new light on the relation between the analytic and the historical approach to various questions in the philosophy of mathematics. Until now these two approaches developed independently, and occasionally there existed tensions between them. One reason for these tensions could be the fact that the proponents of the historical approach to philosophy of mathematics, as for instance Lakatos and some participants of the debate on revolutions in mathematics, such as Crowe or Dauben, understood history primarily as a space for the clash of ideas and for the conflict of opinions, i.e., as an area for social negotiation about norms and values of rational discourse. To logically oriented analytic philosophers, historical reconstructions of such a sociological sort appeared to be missing the (logical) core of the philosophical problems. Therefore they took a negative stance not only on the particular historical reconstructions, but on the very possibility of a historical approach to problems in the philosophy of mathematics.

Nevertheless, our analyses have shown that a broad scale of *historical* changes in mathematics can be studied using categories such as logical or expressive power of language, which are rather close to the *analytic* approach. The point is that historicity must not necessarily mean a historicity of social values and norms. A rather important kind of historicity is the historic nature of our linguistic tools. It would be obviously naïve to believe that a historical reconstruction of the development of the language of mathematics will remove all conflicts between the historical and logical approaches to the philosophy of mathematics. Nevertheless, it could contribute to their rapprochement.

Notions such as *logical, expressive, explanatory*, or *integrative power of* a language are sufficiently exact to be acceptable to an analytic philosopher. In the course of history these aspects of the language of mathematics underwent fundamental changes (as was recognized also by Frege). Thus the historical development of mathematics deals not only with changes of values and norms. There were also changes that are much more accessible to logical description and analysis. The problems posed by development of the different aspects of the language of mathematics could engage also the logically oriented philosopher of mathematics and invite him or her to see in history more than a tangle of psychological and social contingences. On the other hand the fact that, in such important ruptures in the history of mathematics as the birth of algebra or of set theory, the logical aspects of language played such an important role could bring proponents of a historical approach to the philosophy of mathematics to the idea to include among factors influencing the historical process also the different aspects of language. These general theses can be supported by two concrete examples.

1.2.1.1. Revolutions in Mathematics

As most of the reconstructions presented in Chapter 1.1 illustrated the importance of history for understanding of the logical structure of the language of mathematics, I will try to illustrate at least by one example the opposite point, i.e., the importance of logical analysis for the proper understanding of the history of mathematics. For this I have chosen the debate on revolutions in mathematics, initiated by the paper *Ten "laws" concerning patterns of change in the history of mathematics* published by Michael Crowe in 1975. The debate was summarized in the collection *Revolutions in Mathematics* (Gillies 1992). From the ten case studies discussed in the collection as candidates for revolutions in mathematics, five were *re-codings*. Dauben discussed the Pythagorean visualization of number and the birth of Cantorian set theory (Dauben 1984), Mancosu discussed Descartes' discovery of analytic geometry (Mancosu 1992), Grosholz discussed Leibniz (Grosholz 1992), and Gillies discussed the birth of predicate calculus (Gillies 1992a). Let us leave aside for a while the question whether these authors considered the particular re-codings as revolutions in mathematics or not. Let us simply take these papers as an indication of the kind of changes in mathematics which historians consider as sufficiently fundamental to try to explain them using Kuhn's theory of scientific revolutions. We have seen that these changes (whether we consider them revolutions or not) can be reconstructed as changes of the logical, expressive, explanatory, and integrative power of the language of mathematics. This indicates that even if the notion of a scientific revolution is a sociological one, it is correlated with something that can be analyzed by purely analytical means.

When we accept this correlation between revolutions in mathematics and re-codings, it can remind us that two important cases were omitted by the authors of the collection. These two cases were the birth of algebra and the birth of iterative geometry. So we see that a historian can get from the analytic approach a notion of completion and well-arrangement of his material. The logician can draw the historian's attention to certain cases which escaped him and also to arrange the material in a transparent way (for instance according to the growing logical and expressive power of the language). Further, the analytic approach can offer the historian a certain criterion of consistency. If a historian wants to pronounce one or two particular *re-codings* as revolutions (as Dauben did), the obvious question arises, why not also pronounce as revolutions the remaining ones? Of course, the historian has here the final word (whether a change in mathematics is or is not a revolution is a historical and not a logical question). But the analysis of re-codings in terms of the logical, expressive, explanatory, and integrative power of language makes it possible to discuss these historical questions in a more systematic way. Another problem is that two case studies presented in *Revolutions in Mathematics* are changes of a different kind, namely *relativizations* (changes that will be analyzed in Part 2 of the present book): Zheng discusses non-Euclidean geometry (Zheng 1992) and Gray discusses changes of ontology in algebraic number theory (Gray 1992). This leads to the question whether it is acceptable to extend the notion of revolution in mathematics also to the changes of this second kind, and if yes, then whether it would not be more appropriate to differentiate between different kinds of revolutions.

Coming back to the question whether re-codings are genuine revolutions in mathematics, we cannot deny that our reconstruction of the development of mathematics in Chapter 1.1 has an anti-Kuhnian flavor. The regularity of the alternation of symbolic and iconic languages, together with a cumulative growth of the logical, expressive, explanatory, and integrative force of the language of mathematics raises doubts about the presence of incommensurability in the Kuhnian sense in the development of mathematics. The symbolic and the iconic languages cannot be mutually translated one to the other; there is a genuine non-translatability between them. Nevertheless, great fragments of them *can be translated*, thus this non-translatability surely is not incommensurability. Thus instead of the controversies whether there are revolutions in mathematics, it would be perhaps easier first to try to characterize in a precise analytic way the different degrees of non-translatability between its symbolic and iconic languages and so to put these controversies on a more solid basis.

1.2.1.2.   The Historicity of Logic

Robinson's discovery of non-standard analysis is an illustration of the logical power of the language of set theory. Nevertheless, besides its importance for mathematics proper, non-standard analysis is important also from the philosophical point of view. It shows that the rejection or acceptance of certain mathematical methods (in cases, when there is no specific error discovered in them) is determined by the linguistic framework on the background of which we judge these methods. In the times of Cauchy, Dirichlet, and Weierstrass the framework of model theory was not available and so *their* rejection of methods based on manipulations with infinitesimals was justified. Nevertheless, by means of *our* model theory it is possible to give to the methods of Leibniz and Euler solid and convincing foundations and so to rehabilitate them against the criticism of Cauchy, Dirichlet, and Weierstrass. So it is possible that also other mathematical theories, which were in the past rejected, could be brought to new life by means of modern logic and set theory.

A shift in the opposite direction occurred in the case of Euclid, who was considered for a long time the paradigm of logical rigor. Pasch, thanks to new mathematical achievements (in the field of real analysis) discovered that many of Euclid's proofs cannot be correct. If we take the plane consisting of points both co-ordinates of which are rational numbers, we obtain a model of the Euclidean axioms. Nevertheless, some of Euclid's constructions cannot be performed in the case of this plane, because several points used in these constructions do not exist there. But as this model fulfills all the axioms, it is clear that the *existence* of the points used by Euclid in his constructions cannot follow logically from his axioms. Pasch's criticism led to attempts to improve Euclid and to find a system of axioms from which all the propositions that Euclid believed to have proven, would really logically follow. The most successful of these new systems of axioms for Euclidean geometry was Hilbert's *Grundlagen der Geometrie* (Hilbert 1899).

The examples of Pasch and Robinson illustrate the historicity of our view of mathematical rigor. The question whether Euclid or Euler did prove certain propositions or not depends on the linguistic framework on the background of which we interpret their proofs. Before the construction of real numbers it was natural to consider Euclid's proofs as logically rigorous, just as before the advent of non-standard analysis it was natural to consider Euler's methods as problematic. Thus the question whether a proposition is correctly proven is historically conditioned. Nevertheless, this historicity of the notion of logical rigor has nothing to do with historicism or relativism. Before Pasch it was objectively correct to hold Euclid's proof to be rigorous, just as after Robinson it is objectively correct to view Euler's calculations with more respect. The relation between logic and history is an objective one; it has nothing to do with the subjective preferences of mathematicians, historians, or philosophers. The point is that the one side (a particular proof of Euclid or Euler) as well as the other side (the mathematician, historian, or philosopher who interprets the particular proof) are historically situated. They are situated in a particular linguistic framework, which decides whether it is possible to reconstruct Euler's arguments or whether it is possible to find a counterexample to a particular proof of Euclid's. Dedekind's construction of a plane containing only points with rational co-ordinates, just like Robinson's construction of the hyper-real numbers, forced us to change our view of certain periods of the history of mathematics.

The discovery of counterexamples to Euclid's constructions was made possible thanks to the linguistic framework of the theory of real functions (which we included in iterative geometry). As Hintikka and Friedman pointed out, in the times of Euclid there simply did not exist the necessary logical tools that would make it possible to realize that two circles that obviously converge can have nevertheless no point in common (as is the case on the plane with only rational points for circles that intersect in a point with irrational co-ordinates – this point simply does not exist on that plane). Obviously we cannot blame Euclid for not knowing the modern theory of the continuum. Therefore Pasch's criticism and Hilbert's axioms are not so much improvements over Euclid's shortcomings, but rather transpositions of Euclid's theory into a new linguistic background. A similar line of thought is possible also in the case of Euler. Even though non-standard analysis made it possible to vindicate several results obtained by Euler, we must admit that in the linguistic framework of eighteenth century mathematics it was impossible to justify Euler's methods. Therefore the critical reaction of the nineteenth century was fully justified. The fact that some later linguistic framework makes it possible to find some counterexamples, or to prove the consistency of certain methods, does not mean that these counterexamples or consistency proofs were possible (or even in some sense did exist) before the new linguistic framework was introduced. Therefore, at least from the *historical* point of view it seems proper to interpret and evaluate every theory against the background of the linguistic framework in which it was formulated. That means Euclid should be discussed against the background of the framework of synthetic geometry and Euler against the background of the differential and integral calculus.

Nevertheless, from a *philosophical* point of view such a requirement seems to be rather too restrictive. We cannot deny that Pasch and Robinson brought something very important. It would be a great loss to give up the new depth of understanding of the work of Euclid or Euler that was attained by Pasch and Robinson because of some doubtful historical correctness. If we give up the role of a historian who understands himself as a judge that must obey justice, a much more interesting possibility opens up for the history of mathematics. We can fully use the intellectual richness of modern mathematics (as we already did when we characterized the expressive and the logical boundaries of a particular language using the tools of the later periods) in order to see the theories of the past from as many points of view as possible. Thus we can try to do systematically what Robinson did in the case of Euler – to see the possibilities and boundaries of the mathematical theories of the past. It may be the case that also other theories of the past were discarded for the wrong reasons (sometimes even before they were fully elaborated). Modern logic and set theory give the historian strong tools which he or she can use to really understand what happened in history; not only on the social or institutional level, but also from the logical point of view.

1.2.2. Perception of Shape and Motion

In the chapter on iterative geometry we quoted a passage from the book *Fractal Geometry of Nature* (Mandelbrot 1977) in which the author compares the universe of fractals with that of Euclidean geometry. Mandelbrot characterized the Euclidean universe as cold; as a universe in which there is no room for the shape of a cloud, for the coastline of an island, or for the form of the bark of a tree. In contrast to this conception, in iterative geometry clouds, coastlines, or the barks of trees start to be shapes in the strict geometrical sense, i.e., shapes that can be generated by the means of geometric language. If we look at the three iconic languages described in Chapter 1.1 – the language of synthetic, of analytic, and of iterative geometry – we can interpret them as three different ways of grasping the phenomenon of form; as three ways of drawing the boundary between form and the formless.

Synthetic geometry tries to grasp the phenomenon of form by means of static objects such as a circle, a square, a cube, or a cone. To this approach, as Mandelbrot noticed, the form of many natural objects remains hidden. Synthetic geometry is suitable first of all for the planning of artifacts such as buildings, bridges, or dams. The second approach to the phenomenon of form is by analytic geometry, which grasps this phenomenon by means of a co-ordinate system and some analytic formulas (polynomials, infinite series, differential equations, and the like). This approach broadened the realm of objects and processes, the form of which can be constructed geometrically. So for instance in the suspended chain we discovered the catenarian curve; in the trajectory of an arrow we found a parabola; in the vibrating membrane we detected Bessel's functions. Even these few illustrations indicate that analytic geometry brought us much nearer to nature. Everywhere where the form is the result of a simple law or of a small number of determining factors, analytic geometry is able to describe it and so make it accessible to further investigation. Nevertheless, Mandelbrot's criticism still remains valid. Objects that are not created at once but which are the result of erosive or evolutionary processes (as the relief of a mountain or the form of a tree), objects which are the result not of one or ten determining factors but of millions of random influences (as a coastline or a cloud), remain formless even from the point of view of analytic geometry. The third approach to the phenomenon of form is by iterative geometry, which grasps this phenomenon as a limit of successive iterations of a transformation. Because the growth of plants and animals happens due to repeated cell division and the different processes of erosion are the result of repeated periodic influences of the environment, iterative geometry is able to grasp many of the natural forms that emerge through growth or erosion. The development along the line of synthetic, analytic, and fractal geometry changed thus the way of our perception of form and shifted the boundary between form and the formless.[20]

From the philosophical point of view this development is interesting, because at first glance it might appear that the perception of form, as well as the boundary between what has form and what we perceive as formless, is something given by our biological or cognitive make-up. Even though I do not deny the importance of biological and psychological factors in perception, it is interesting to notice, that perception has also a linguistic dimension. Changes of the language of mathematics; the birth of analytic and of iterative geometry made it possible to create a radically new universe of forms. These new forms appeared first in the minds of a few mathematicians, later they entered into the theories of physicists and other scientists, and finally, due to the progress of technology, the new forms percolated also into the socially constructed reality of our everyday lives. Thus mathematics also belongs to the factors that influence our perceptional function. Anyone who has studied fractal geometry perceives the leaf of a fern or the shape of a mountain in a new way. Therefore it is probable that modern man perceives forms differently from the ancient Greeks, and that at least part of this difference in perception is caused by mathematics. It is probable that mathematics shapes not only the way we think, but it determines also how we perceive the boundary between form and the formless.

If we look from this point of view at Aristotle's theory of natural motions, the fact that he recognized only rectilinear and circular motions as being natural, can be interpreted as the "influence" of Euclid.[21] In Euclidean geometry the circle and the straight line are privileged forms, and Aristotle in his theory of natural motion simply repeats this geometrical distinction. Thus Aristotle found in nature rectilinear and circular motions not for some physical reasons, and not even because of some metaphysical preferences, but he simply copied them from ge-

---

[20] It is interesting that in ordinary language there is no word for the opposite of formless; a word that would signify all objects that have form. It seems that the language perceives form as something that constitutes the object as such; as something that the object cannot be deprived of. It seems as if form is not an ordinary predicate (like color). It seems as if form would belong to the essence of an object, and therefore we need no special word to indicate that objects have form. It is sufficient that we have a word for existence, because to exist and to have form is the same.

[21] Of course I am aware that Euclid lived after Aristotle, therefore if there was an influence, it had to have the opposite direction. Nevertheless, Euclid's *Elements* are a culmination of a particular tradition. When I speak about Euclid's influence on Aristotle, I have in mind the influence of this tradition (represented by Eudoxus, Theaetetus and other mathematicians whose discoveries are contained in the *Elements*). Euclid can be seen as the embodiment of this tradition.

ometry. When in the seventeenth century analytic geometry opened the door into a new universe of forms, the circle and the straight line lost their privileged position. At the same time physicists started to se-lect the trajectories of mechanical motions from a much wider range of possible curves. They discovered that a stone thrown in the air follows a parabola, that the planets orbit the sun on ellipses, and that the tra-jectory of fastest descent (the brachystochrona) is a cycloid. Against the background of this broader universe of forms Aristotle's theory appeared as an artificial restriction of the possible trajectories. The view that rectilinear and circular motions are natural became obscure. These motions were natural only against the background of Euclidean geometry, but against the background of analytic geometry they are not more natural than any other polynomial curve of a low degree. Sim-ilarly when in the nineteenth century the first fractals appeared, they slowly found their way into physics (among their first application was the theory of Brownian motion developed by Norbert Wiener). Thus the different geometrical languages were important not only from the point of view of the perception of *form*, but because they influenced in a fundamental way also our perception of *motion*. Just like geometrical shapes, the different physical processes can also be represented using synthetic, analytic, or iterative geometry.

Perhaps the most interesting aspect of our reconstruction of the de-velopment of the iconic languages is that this development is not con-fined to the world of geometry. The transitions from the synthetic through analytic to fractal geometry did not happen inside of the ge-ometric universe. The new geometrical languages appeared always thanks to a symbolic intermediate level. Descartes was able to break the narrow barriers of the Euclidean universe and to open the door into the universe of analytic curves only thanks to the language of algebra. Sim-ilarly for the discovery of the universe of fractal geometry the language of the differential and integral calculus played a decisive role. This in-terplay between the development of geometry and the development of symbolic languages sheds new light on philosophical theories that base mathematics on intuition (Kant's transcendental idealism or Husserl's transcendental phenomenology). These philosophies can deepen our understanding of the visual aspect of mathematics; they can explain how our perception of shape is constituted and what its structure is. Nevertheless, they are unable to understand its changes, because the changes of our perception of form do not take place in the geometrical realm alone. Thus unless the philosophy of intuition is completed by a theory of symbolic languages, it cannot understand the differences between intuition in synthetic and in analytic or fractal geometry. Therefore such philosophies can at best produce a philosophical reflection of the Euclidean geometry, but they cannot be extended into a philosophy of the whole realm of geometry, let alone of the whole of mathematics.

1.2.3. Epistemic Tension and the Dynamics of the Development of Mathematics

Every period in the history of mathematics (with the exception of ancient Egypt and Babylonia) has a symbolic and an iconic language. These two languages determine the universe of the mathematics of the respective period. For instance the mathematics of the Renaissance was based on the symbolic language of algebra and the iconic language of synthetic geometry. An interesting aspect of these two languages is their untranslatability. The iconic and the symbolic languages are never coextensive. There is always an *epistemic overlap* between them. The diagram on page 86 makes it possible to determine this overlap more precisely. The language which is placed lower in the table (i.e., which is historically younger) is stronger than the language that is placed higher (i.e., which is older). Thus in the case of the Renaissance the symbolic language of algebra had an *epistemic overlap* over the iconic language of synthetic geometry. This epistemic overlap of the language of algebra over the language of synthetic geometry means that in the language of algebra there were expressions which could not be interpreted geometrically. Therefore in the Renaissance it was possible to calculate more than it was possible to represent geometrically.

Such a situation causes a tension; it creates a need for change of the iconic language so that it becomes possible to find some geometrical representation also for those algebraic expressions for which synthetic geometry is unable to give an interpretation (such as the fourth power of the unknown). When Descartes created analytic geometry, which made it possible to represent geometrically any power of the unknown, a new epistemic overlap occurred, but this time on the side of the iconic language. Analytic geometry enables us to draw not only curves that are defined by means of a polynomial, but also curves such as the logarithmic curve or the goniometric curves, i.e., curves which cannot be represented algebraically. This epistemic overlap on the iconic side was reduced by Leibniz who laid the foundations of the differential and integral calculus. But as it later turned out, the new symbolic language based on the notion of the limit transition had an epistemic overlap over the language of analytic geometry. In order to reduce this epistemic overlap, iterative geometry was created.

This shows that the epistemic overlap is irreducible. A perfect harmony between the symbolic and the iconic poles of mathematics seems to be impossible. And precisely the irreducibility of the epistemic tension between the symbolic and the iconic languages is the basis of our reconstruction of the development of mathematics. We call the diagram, which we presented on page 86 as a summary of our reconstruction, the *bipolar diagram*. It represents the development of mathematics as a process of oscillations between two poles. In our view it is a more appropriate scheme than the classical view of the development of mathematics as a cumulative process. It captures not only the growth and differentiation of knowledge (which can be obtained from our scheme if we restrict ourselves to one of the poles, as for instance Frege, who described only the development of the symbolic pole), but it captures also the epistemic tension, which drives this growth. The two poles, even though they are mutually irreducible (it is impossible to construct a symbolic language that would have *exactly* the same logical, expressive, explanatory, and integrative power as a particular iconic language), do not exclude each other. On the contrary, they complement each other and together provide mathematics with the necessary expressive and logical means. The tension between the two poles drives mathematics to create always new symbolic and iconic languages. For the description of the relation of these two poles perhaps Bohr's notion of *complementarity* is most suitable.

1.2.4. Technology and the Coordination of Activities

We have shown that the development of geometry influenced our perception of form. The question arises whether also the development of the symbolic languages of mathematics from elementary arithmetic and algebra to the predicate calculus changes somehow our perception of our surroundings. Unfortunately, it is not easy to answer this question unequivocally. In the case of geometry it was clear from the very beginning that all three iconic languages have something to do with the perception of form. On the other hand in the case of the symbolic languages it is not clear what that something should be, the perception of which is changing when we pass from arithmetic to algebra. Therefore first of all we have to answer the question what should be the analogy of the phenomenon of form, on the changing perception of which we could follow the development of the symbolic language of mathematics. One of the possibilities is to turn to technology. The algebraic symbols were created in order to be able to represent particular (mathematical) operations. Thus the levels in the development of the symbolic languages could correspond to changes of the principles of coordination of human activities, and thus with the fundamental changes of technology. Nevertheless, this correspondence between the symbolic languages and types of technology is much more tentative than the one between iconic languages and the perception of form.

The *handicraft technology* of Antiquity and the Middle Ages can be interpreted as the technology that is based on the same schemes of coordination, on which the calculative procedures of elementary arithmetic were based. The craftsman manipulates with particular objects similarly as the Egyptian reckoner manipulated with numbers following the instructions of the Rhind papyrus. Thus, just like Frege characterized elementary arithmetic, we can say about handicraft technology that it consists in the manipulation of constant objects.

Machine technology consists in dividing the technological process into its elementary components and letting each worker perform only one or a few these elementary operations. Thus the technological process is divided into its elementary components just like an algebraic calculation is divided into elementary steps which form the constituents of an algebraic formula. Similarly as the formula represents a particular numerical quantity (for instance the root of an equation) in the form of a series of algebraic operations, the technological process "represents" (or produces) the particular product in a series of technological operations which gained a relative independence and thus can be performed by different workmen. The generality which in algebra is due to the use of variables, is present in the technological process in that a given workman performs the particular technological operation in its generality, i.e., with all objects that pass him on the production line. In contrast with the craftsman, who makes all the operations of the technological process with one object from the beginning to the end (and so his operations are manipulations with that constant object), the workman performs his particular operation with all objects (and so his object is "variable"). Thus just as Frege characterized the language of algebra, we can say about machine technology that it consists of manipulation with variables (or variable objects).

The technology with schemes analogous to the structure of the language of differential and integral calculus could be *chemical technology*. The technological processes in chemistry do not have the form of a series of separate elementary operations, which have to be performed one after the other, but rather it is a continuous process that has to be controlled. Mixing some reagents is not an operation parallel to those that we know from algebra. Mixing, heating, or adding some reagents are continuous processes by means of which we can control the chemical reactions.[22] Thus in analogy with Frege's characterization of the language of the calculus, we can also say about the technological processes in chemistry that they consist of manipulations with functions of second degree.

When we look for technologies that would be analogous to the language of the predicate calculus we can take the *analog controlled technology*. The control by means of an analog computer makes it possible to realize almost any technological process. When we construct the respective logical circuit, it will control the technological process with the required precision. Thus in analogy to Frege's characterization of the language of the predicate calculus we can say also about the schemes of analog control in technology that they consist of manipulations with arbitrary second-level functions.

This short account does not aspire to be a universal history of technology. Its aim was rather to offer a new view on the nature of technological innovations. The history of technology is usually brought into connection with the development of physics. This, of course, is correct. Nevertheless, we would like to call attention to the fact that many changes in the general structure of technology can have close relations also with the development of mathematics. The creation and spread of algebraic symbolism in western mathematics happened shortly before the birth of machine technology. It is possible that this was not a mere coincidence. The symbolic thought that was cultivated in algebra could

---

[22] If the analogy between the differential and integral calculus and chemical technology is legitimate, it opens a new perspective on one peculiar aspect of Newton's works—Newton's alchemy. Newton's works on alchemy were for a long period ignored by historians. John Maynard Keynes made the first attempt to understand this part of Newton's work (Keynes 1947). If the interpretation of chemical technology as the area in which the character of coordination of activities is analogous to the character of operations in differential and integral calculus is correct, then Newton's penchant for alchemy comes into new light. It is possible that between alchemy and the differential and integral calculus there was a deep analogy that fascinated Newton. Thus his predilection for alchemy was not just an extravagancy of a genius but maybe had a rational core.

contribute to a change in the perception of technological process. And it cannot be excluded that the development of the symbolic languages of mathematics influenced the advancement of technology also in other cases. Mathematics offers tools which enable us to perceive the coordination of technological operations in an efficient way. Therefore one of the contributions of mathematics to the development of western society could lie in the cultivation of our perception of algorithms.

1.2.5. The Pre-History of Mathematical Theories

When we look at the development of a particular mathematical discipline from the point of view of its content (and not its language, as we did in Chapter 1.1), i.e., when we look at the concepts, methods, and propositions that form this discipline, we will find that mathematicians discovered the main facts about this new domain *long before an adequate language was created*, which enabled them to express these findings in a precise way and to prove them. Perhaps the best illustration of this is the integral calculus. The first mathematical results in the field that later was called integral calculus were achieved by Archimedes almost two thousand years before Newton and Leibniz created the language which allowed one to calculate integrals by means of purely formal operations. Archimedes worked in the framework of synthetic geometry, which is not suitable for the calculation of areas and volumes of curvilinear objects. Therefore he used the mechanical model of a lever in his calculations as a heuristic tool. From the two arms of the lever he suspended parts of different geometrical objects and then from the conditions of equilibrium of the lever he was able to derive relations among the areas or volumes of the suspended objects. The main disadvantage of this method consisted in the fact that for each geometrical object he had to invent an ingenious way of cutting it into pieces and of balancing them by other geometrical figures in order to obtain equilibrium on the lever. Thus the greatest part of Archimedes's ingenuity was consumed by the difficulties posed by the language of synthetic geometry that is unsuitable for the calculation of areas and volumes of curvilinear figures.

Every further language in the development of mathematics enriched the arsenal of methods that were at disposal for the calculations of areas and volumes. Thus the birth of algebraic symbolism made it possible to generalize the calculations of the area of a square and of the volume of a cube to the arbitrary power x^n. Even though this problem lacked any geometrical interpretation, the language of algebra enabled Cavalieri using his method of *"summing the powers of lines"* to find the relations, which correspond to our integrals $\int_0^a x^n dx = \frac{1}{n+1} a^{n+1}$ for $n = 1, 2, 3, \ldots, 9$ (Edwards 1979, p. 106–109). For instance for $n = 5$ he formulated his result as: *"all the quadrato-cubes have to be in the ratio of 6 : 1"*. The *quadrato-cubes* represented the fifth powers of the unknown and the ratio 6:1 gives the reciprocal value of the fraction that stands on the right-hand side of our formula for the integral of x^5. Even though it was not clear what he was calculating, the results of Cavalieri were absolutely correct.

The birth of analytic geometry brought a new interpretation of algebraic operations. Thus x^3 was interpreted not as a cube, as this expression was interpreted by Cavalieri, but simply as a line segment of the length of x^3. This change of interpretation together with the idea of a co-ordinate system enabled Fermat to find a new method for calculating Cavalieri's cubatures. Fermat did not sum *"powers of lines"*, and his cubature was not a calculation of the volume of some four-dimensional object, as it was for Cavalieri. Analytic geometry enabled Fermat to interpret this cubature as the calculation of the area below the curve $y = x^3$. Thanks to a suitable division of the interval $(0, a)$ by points forming a geometric progression, this integral is easy to find. And perhaps what is even more interesting, Fermat's method is universal, it can be made with n as parameter. Thus while Cavalieri had to find for each case a special trick to sum his powers, the method of dividing the interval by points that form a geometric progression, works uniformly for all n. This example nicely illustrates the increase of the integrative power of the language.

The growth of the expressive and integrative power of language by passing from synthetic geometry, which was used by Archimedes, through algebra that was employed by Cavalieri, to analytic geometry used by Fermat, enriched in a fundamental way the methods of integration. This enrichment enabled Newton and Leibniz to discover a fundamental unity of all these methods and to incorporate it into the syntax of the newly created integral calculus. The creation of a language that makes it possible by means of manipulation with symbols to solve the problem of quadratures and cubatures was a decisive turn in the development of mathematics. From the point of view of the integral calculus all results obtained by mathematicians of the past seem to be no more than simple exercises. The tricks of an Archimedes, Cavalieri, or Fermat are impressive, but a second-year university student manages in a few hours what represented the apex of their scientific achievements.

In mathematics education this organic growth of linguistic tools is often ignored and the teaching of a particular mathematical discipline starts with the *introduction of the language*, i.e., with the introduction of the symbols and syntactic rules by means of which it became possible to formulate the discipline's basic notions and results in a precise way. So the teaching of algebra starts with the introduction of the symbol x for the unknown, and mathematical analysis starts with the introduction of the syntactic distinction between function and argument. Thus the students do not develop their own cognitive process of formalization; they do not learn how to formalize their own intuitive concepts and how to create a symbolic representation for them. Only if we keep in mind the centuries of gradual changes and the dozens of innovations that separate the quadratures of Archimedes from the brilliant technique of integration, say of Euler, can we realize the complexity of the difficulties involved in mastering the linguistic means used in a mathematical domain such as the calculus. The historical reconstructions show that it is inappropriate to start the teaching of a mathematical discipline by the introduction of its linguistic tools. Only when the linguistic tools are brought into a relation with the cognitive contents that they have to represent, only then can they be adequately mastered.

When we write a formula, we are usually unaware of the centuries of mathematical experience that are present in it in a condensed form. Only when we dissect the language into its historical layers does it become obvious how many innovations must have taken place in order to make it possible to write for instance the principle of mathematical induction

$$\left\{\varphi(0) \wedge (\forall n)\big[\varphi(n) \Rightarrow \varphi(n+1)\big]\right\} \Rightarrow (\forall n)\varphi(n).$$

Here the Cossist invention of the representation of the unknown by a letter (in this case with the letter n), is combined with Leibniz's distinction between a function and an argument (in this case $\varphi(n)$), and Frege's invention of the quantification of parts of a formula. And here we mention only the most important of these innovations, because for instance Leibniz's distinction between a function and an argument was the consummation of a long process starting with Archimedes (a few moments of this process, connected with Cavalieri and Fermat, were touched on above). Only if we keep in mind all these changes, can we appreciate the problems that mathematical formalism presents to our students.

CHAPTER 2

Relativizations as the Second Pattern of Change in Mathematics

---

We based our description of re-codings in the history of mathematics on Frege's interpretation of the development of arithmetic as a gradual growth of the generality of its language. Frege identified as the fundamental events in the history of mathematics the invention of the *constant symbols* in arithmetic, the introduction of the *variable* in algebra, the introduction of *symbols designating functions* in mathematical analysis, and finally the introduction of symbols for *functions of higher orders* in logic. When we complemented this interpretation of the development of the symbolic language of mathematics by an analogous interpretation of the development of the language of geometry, we obtained the first pattern of change in the development of mathematics. Nevertheless, it is important to realize, that this union of arithmetic with geometry contradicts Frege's original intention. In arithmetic Frege endorsed the logicist view, according to which the propositions of arithmetic are analytic and can be derived from logic. On the other hand, in the field of geometry he accepted Kant's philosophy according to which the propositions of geometry are synthetic and are based on intuition (see Frege 1884, pp. 101–102). Therefore Frege would probably view our attempt to put arithmetic and geometry into a single linguistic framework in order to study the transitions between them as mistaken. Despite this I believe that the relations between the symbolic and the iconic languages, which our analyses discovered, vindicate this unfaithfulness to Frege.

It is interesting that also for the second pattern of change in the development of mathematics, which is the subject of this section of the present book, it was possible to find a theoretical framework for their analysis. This framework is Wittgenstein's picture theory of meaning from the *Tractatus* (Wittgenstein 1921). But just as in the case of Frege, whose views were used in the first part of this book, so also in the case of Wittgenstein we will only take a few ideas from him and employ them as a tool for the interpretation of the development of mathematics. We will be forced to ignore the rest of his philosophy together with its general intention, just as in the case of Frege. In Frege we had to ignore the sharp division line that he put between the symbolic and the iconic languages of mathematics. Only when we did so, did a unified pattern of re-codings in mathematics emerge. In the case of the picture theory of meaning we will ignore Wittgenstein's thesis about the existence of a single pictorial form common to all languages (a thesis which is the core of the whole *Tractatus*). Only when we accept the idea that the language of geometry or algebra is gradually passing through stages which differ in their pictorial form, will we be able to use (a fragment of) Wittgenstein's picture theory of meaning as a tool for analysis of the subtle semantic shifts that occurred in the development of these mathematical disciplines.

The philosophy of early Wittgenstein relies on two principles. The first is the thesis that *language functions like a picture*. This means that beside logic and grammar there is a further structure of language, independent of the first two, which he called the pictorial form. The second principle is the thesis that *there is only one pictorial form*, common to all languages. Later Wittgenstein abandoned the radical position of his early writings and developed the theory of language games. He had good reasons for abandoning of the *Tractatus*. The picture theory of meaning is too restrictive, if we want to use it in order to understand ordinary languages (such as English or German). But the picture theory of meaning contains several important insights which can be useful for understanding the semantic structure of the languages of mathematical theories. What we have to do is just to liberate the picture theory from the thesis of the existence of unique pictorial form.

What follows is not an attempt to reconstruct the historical Wittgenstein. I will take only a few aspects of the picture theory of meaning and use them as an interpretative tool that makes it possible to recon-

struct the development of the semantic structure of languages of different mathematical theories. Many important changes in the history of
geometry or algebra can be understood if we interpret the development
of these disciplines as the development of the pictorial form of their
language. In order to do this we have to change the *Tractatus'* theory
(and terminology) and to accept two principles:

1. *The existence of the form of language*[1] as a structure encompass-
   ing all that cannot be explicitly expressed in the language but is
   only shown.

2. *The plurality of the forms of language* – the language at every
   stage of its development has only one form, but this form can
   vary in time.

   The concept of the form of language may be important for understanding of the development of mathematics. It is so, because this concept indicates that besides everything that can be explicitly expressed in
a language (and was therefore in the limelight of history of mathematics), there is an implicit dimension of every language that comprises
everything that can be only shown but not expressed by it. It seems
that in the development of mathematics this implicit component played
an important role, which, nonetheless, has not been sufficiently understood, because of the lack of appropriate theoretical tools for its study.
The picture theory of meaning can direct our attention to the study of
the implicit aspects of mathematics. If we allow a plurality of forms
of language, it will become possible that the *form of the language* J_1,
which cannot be expressed in the language J_1 itself, will be *expressible
in a language* J_2. In this way the language J_2 can serve as a tool for an
explicit expression of the form of the language J_1. This would make
reconstruction of the development of the form of language (for instance
of geometry) a well defined and manageable task.

---

[1] Wittgenstein on the *Tractatus* believed that there is only one pictorial form. Therefore it was not
necessary to further specify this notion. In the text that follows we hold the view that there are several
different pictorial forms, i.e., several ways in which language can picture the world. Further we believe
that many of the fundamental changes in the development of a particular mathematical discipline
are connected with the transition from one pictorial form to another. As we will distinguish several
pictorial forms, we have to specify this notion. Therefore we will bind the notion of a pictorial form
to a *particular language* (or mathematical theory, such as algebra or synthetic geometry) and to a
particular stage of its development. In order not to cause confusion, we will for this – history and
theory bounded – notion of form use not Wittgenstein's term of pictorial form; but rather we introduce
for it a new technical term, namely *form of language*.

It turns out that such an explicit expression of the form of the language J_1 by means of the language J_2 happened in the history of geometry several times. This is exactly what was done by Beltrami with the form of the language of Lobachevski's non-Euclidean geometry; by Klein with the form of the language of Cayley's theory of the metric structure of the non-Euclidean plane; or by Poincaré with the form of the language of Riemann's classification of surfaces. And further, incorporation of the form of language J_1 explicitly in the language J_2 opens up the possibility of the *emergence of a new form of language*. Thus if we liberate the picture theory of meaning from the thesis of the existence of only one form of language, we obtain a tool which makes it possible to describe the evolution of mathematical theories. The evolution consists in two alternating processes – the explicit incorporation of the form of language which was at the previous stage only implicit, and the emergence of a new implicit form in the place of the previous one which was made explicit.[2]

The usefulness of the concept of the form of language is based on two circumstances. Firstly this concept is closely related to the notion of the subject (*Tractatus* 5.632: "*The subject does not belong to the world: rather, it is a limit of the world*") and therefore it makes it possible to reconstruct such things as translation between languages or understanding of a language without the need for introducing a subject from outside in the form of an idealized scientist or a scientific community. We need not to introduce a subject from outside (from sociology as in Kuhn, or from psychology as in Piaget), because the subject is already present as a constituent of the form of language. That means that we are not forced to mix our epistemological considerations with sociological or psychological elements. And secondly the form of language is clearly separated from logic. Therefore the changes of the form of language in the development of a mathematical discipline do not interfere with logical consistency.

In this chapter we will describe the changes that occurred in synthetic geometry on the road to non-Euclidean geometries. We will follow this developmental line till Hilbert's axiomatic system. The reconstruction of the development of geometry will be followed by a similar

---

[2]   The possibility of expressing in an explicit way the form of language J_1 in the language J_2 opens the possibility for the emergence of a series of closely connected languages, where the stage J_n would contain in an explicit way the form of language of the preceding stage J_{n-1}, while its own form of language would become explicit only in the next stage J_{n+1}.

reconstruction of algebra which will be centered around the historical development leading to Galois theory. In the last sections of this chapter we compare the development of geometry with the development of algebra and we will try to reach a coherent picture of this second pattern of change in the development of mathematics, which we will call *relativizations.*

2.1. Historical Description of Relativizations in Synthetic Geometry

The discovery of non-Euclidean geometries is one of the most often discussed events in the history of mathematics. Some authors are convinced that several aspects of non-Euclidean geometry were known already in Antiquity. These authors consider the complicated form which Euclid gave his fifth postulate to be the result of discussions of several alternative geometrical systems that are lost (see Tóth 1977). Nevertheless, the majority of historians of mathematics put the date of the birth of non-Euclidean geometry at the end of the eighteenth and the beginning of the nineteenth centuries (see Bonola 1906, Kagan 1949, Kline 1972, Rozenfeld 1976, Gray 1979, Boi, Flament and Salanskis 1992, Boi 1995). But whether we consider the beginning of the nineteenth century as the date of the discovery or of the rediscovery of non-Euclidean geometry, the fact remains that from the fourth century A. D., when Proclus pronounced his doubts about the fifth postulate till the beginning of the nineteenth century, when Gauss, Bolyai, Lobachevski, Taurinus, and Schweikart created their non-Euclidean systems a *very long time* passed. It seems as if something was hindering mathematicians from entering the world of non-Euclidean geometry, as if there were some barrier that at the beginning of the nineteenth century suddenly fell and opened the way to these new geometries. But the fifteen centuries that separate the formulation of the problem of the fifth postulate from its solution is not the only peculiarity of this discovery. Another remarkable circumstance is its *high degree of parallelism.* After fifteen centuries of unsuccessful attempts suddenly, in the course of a few decades, five mathematicians came independently to similar results. Igor Shafarevitch formulated this aspect of the discovery of non-Euclidean geometry as:

> "After Lobachevski and Bolyai laid the foundations of non-Euclidean geometry independently of one another, it be-

came known that two other men, Gauss and Schweikart, also working independently, had come to the same results ten years before. One is overwhelmed by a curious feeling when one sees the same designs as if drawn by a single hand in the work done by four scientists quite independently of one another." (Davis and Hersh 1983, p. 53)

I will try to interpret the discovery of non-Euclidean geometries as a result of a series of linguistic innovations. Thus we will concentrate our attention not only on what mathematicians such as Lobachevski or Bolyai discovered, but we will analyze also the tools by means of which they expressed their discoveries. The analysis of the linguistic innovations that accompanied the discovery of non-Euclidean geometry will shed light on the remarkable circumstances of this discovery.

It will be possible to explain the *"existence of non-Euclidean systems before Euclid"* discussed by Tóth by the circumstance that as long as the linguistic framework, which lies at the basis of the Euclidean system, had not stabilized, it was possible to discuss many propositions and arguments that contradict this framework. Establishment of the linguistic framework in which Euclid's elements are formed excluded these propositions and arguments as inconsistent or even as incomprehensible. So the history of mathematics is not a simple cumulative process of collecting knowledge. If we analyze the linguistic tools by means of which mathematical knowledge is formulated, it will be possible to show that the establishment of a particular linguistic framework leads to an easier and more accurate formulation of certain propositions. But at the same time the new framework hampers or even precludes the formulation of many other propositions, which start to be considered as obsolete or even inconsistent, i.e., as not reaching the standards of precision of the new framework. An example of such a rejection of a whole mathematical theory was the discarding of infinitesimals after introduction of the framework of $\varepsilon - \delta$ analysis by Weierstrass. It is possible that the "forgetting" of the non-Euclidean systems that existed before Euclid was a further illustration of this phenomenon.

The long time that passed between expression of doubts about the fifth postulate and discovery of the first non-Euclidean systems can be explained when we take into account the linguistic innovations that made this discovery possible. It turns out that the linguistic framework in which a consistent system of non-Euclidean geometry can be formulated differs from the framework of the Euclidean system not by one but by a whole series of linguistic innovations. In the present chapter we will describe the most important of them. I am convinced that, in order to be able to create a consistent system of non-Euclidean geometry, it was necessary to introduce into geometry the notion of *space*, the notion of a *projection*, and at least an implicit notion of a *model*. These all are tools that Lobachevski used in the construction of his system, but which cannot be found in Euclid. Euclid did not have the notion of space; he spoke only about objects, i.e., about what there is, but never about the empty space that surrounded these objects (see Kvasz 2004). For non-Euclidean geometry the notion of space is very important because the simplest way of envisioning a non-Euclidean geometry is to imagine it as the geometry of a non-Euclidean *space*. Similarly, in his geometry Euclid did not use projections. Nevertheless, the decisive step of Lobachevski was the *projection* of the non-Euclidean plane onto a surface, called a horosphere, which served as a *model* (in the non-Euclidean space) of the Euclidean plane. So we see that all three notions – space, projection, and model – played an important role in the system of Lobachevski. Nevertheless, the introduction of the notions of space, projection, and model into geometry did not happen at once. It took centuries until linguistic frameworks were developed in which it was possible to work with these notions in a consistent manner. The time that had to pass until these frameworks were sufficiently mastered can be seen as an explanation of the long time we had to wait for the first non-Euclidean systems.

We will try to solve the problem formulated by Shafarevitch in that the hands that *"drew the designs in the work done by four scientists quite independently of one another"* were led by common form of language. We are convinced that Bolyai, Lobachevski, Gauss, Taurinus and Schweikart introduced a particular linguistic innovation that enabled them to express in a consistent way several propositions of non-Euclidean geometry. The high level of mutual resemblance among the designs contained in their works can be explained by the fact that different constituents of the language of a mathematical theory are strongly interconnected. Therefore there is only a very small number, and in some cases only one linguistic innovation that would not lead to logical contradictions. This makes it probable that different mathematicians, working independently of one another, would make the same innovation which will consequently lead them to analogous results.

2.1.1.   The Perspectivist Form of Language
of Synthetic Geometry

One of the interesting aspects of ancient mathematics was that it did not know the notion of space. The closest Ancient notion to what we today call space was the concept of emptiness ($\kappa\varepsilon\nu o\nu$). But to many Ancient thinkers, perhaps with the exception of the atomists and Epicureans, the notion of emptiness seemed to be problematic. Emptiness is where there is nothing and so this notion designates something which cannot have any specific attributes that we could study. Even the atomists, who admitted the existence of emptiness, were unable to say anything particular about it. And independently of what we think about the existence of emptiness, this concept surely could not be the subject of mathematical investigation. Mathematics was always the paradigm of lucidity and exactness, but it was impossible to reach any lucid and exact knowledge about emptiness. Maybe this was the reason why the concept of space was not created in mathematics but in art.

2.1.1.1.   *The Implicit Variant of the Perspectivist Form – Giotto and Lorenzetti*

If we compare the paintings of the Renaissance with the paintings of the preceding period, we immediately notice a striking difference. Gothic paintings lack depth. The figures are placed beside one another, house beside house, hill beside hill, without any attempt to capture the depth of the space. In handbooks on Gothic painting, we can find the explanation for this. This kind of painting was in agreement with the general aims of the painter. The painter's task was not to paint the world as it appeared to him. He had to paint it as it really was, to paint it as it appeared to God. The distant objects appear to us smaller, but they only appear so, in reality they are not smaller at all. So the painter must not paint them smaller

One of the first painters in whom we can recognize a systematic effort to capture the depth of space was the Florentine painter Giotto di Bondone. In most histories of art his work is considered to be one of the supreme accomplishments of medieval art, and it cannot be denied that his use of themes as well as iconography set him well within the medieval period. On the other hand, we can find in his paintings features which are typical of Renaissance art. These features are related to the geometrical aspect of his paintings, in which a discernible effort to represent real geometrical space and the arrangement of objects in this space can be seen. Because the subject of our interest is precisely those geometrical aspects of painting in which Giotto was well ahead of his times, we have chosen his fresco *Revelation to Father Augustin and the Bishop* (around 1325) from Santa Croce in Florence as our first illustration of the principles of Renaissance art. As we can see from the auxiliary lines, complemented by Kaderavek, Giotto already understood that in order to evoke the illusion of parallel lines he had to paint lines which converge to a common point (see Kaderavek 1922). Yet in the fresco the point of convergence of lines belonging to the ceiling and the point of convergence of lines belonging to the canopy are different. It is clear from the painting that there are two main points (points of

convergence of lines leading to the depth of space) and also two horizons. That Giotto used different main points and different horizons means that the perspective accomplished in his fresco is rather "fragmentary", that is, different parts of the architecture are painted from different points of view.

The next painting with which we would like to illustrate the discovery of the principles of perspective comes from Ambrogio Lorenzetti and is called *Annunciation* (1344). From the geometrical point of view there is a very interesting representation of the pavement in the painting. Lorenzetti uses the pavement to evoke an illusion of the depth of space. The lateral sides of the tiles forming the pavement converge to a unique point, which is thus the main point of the painting. This

construction of the painting is in accordance with the principles of perspective. Nevertheless, if we draw into the pavement the diagonals of the tiles, we will discover that they form a curve. From the geometrical point of view, the perspective is false, as in reality the diagonals form straight lines, and in the central projection, the image of a straight line is always a straight line. Accordingly the line formed by the diagonals of the particular tiles of the pavements should be a straight line. Thus the question of the correct representation of pavement posed an important problem. We see that Lorenzetti already knows that the tiles of the pavement should gradually diminish, but he fails to surmise the precise ratio of this diminishing.

2.1.1.2.   *The Explicit Variant of the Perspectivist Form – Masaccio*

One of the first paintings the spatial structure of which was geometrically constructed was the fresco *The Saint Trinity* (1427) in the church Santa Maria Novella in Florence. Its author was Tommaso di Set Giovanni di Simone Cassai, called Masaccio. Thus the basic principles of perspective, which we have seen by Giotto and Lorenzetti in an intuitive form, were by Masaccio fully mastered. Masaccio explicitly knew and precisely applied all the rules that are necessary to create an illusion of the depth of space. Therefore we can say that the perspectivist form of language became here fully explicit. The Renaissance painters

wanted to paint the world as they saw it, to paint it from a particular *point of view*, to paint it in perspective. They wanted to paint the objects in such a way that the picture would evoke in the spectator the same impression as if he was looking at the real object. Thus, it had to evoke the illusion of depth.[3] To reach this goal the painter followed the principles of perspective:

---

[3] The passing of parallels into the depth of the space, just like the depth of the space itself, belong to the form of the picture. A planar picture cannot express depth; depth is only shown by the picture. Therefore we must learn to see the depth of space in perspectivist pictures. A picture is a planar structure and as such it has no depth. What is explicitly present in the picture, and therefore can be seen by anyone, is that particular lines converge. Nevertheless, despite the planar nature of pictures some of them can show depth of space.

Perspective of size – the remote objects are to be painted smaller. *Perspective of colors* – the remote objects are to be painted with dimmer colors. *Perspective of outlines* – the remote objects are to be painted with softer outlines

The first of these is connected with space; the other two are caused by absorption of light in air.

By following these principles a special line appears on the painting – the *horizon*. In fact the painter is not allowed to create it by a stroke of his brush. He is not permitted to paint the horizon, which shows itself only when the picture is completed. According to proposition 2.172 of the *Tractatus* ("*A picture cannot, however, depict its pictorial form: it displays it.*"), the horizon belongs to the form of the language. It corresponds to the boundary of the world pictured by the painting, and therefore, according to proposition 5.632 ("*The subject does not belong to the world: rather, it is a limit of the world*"), the horizon belongs to the subject. So besides the signs of the iconic language which express definite objects, there are expressions on the painting connected not with the objects, but with the subject, which is the bearer of the language.

2.1.2. The Projective Form of Language of Synthetic Geometry

The construction of perspective, especially in the case of the diminishing tiles of a pavement, is a rather complicated procedure. Many painters tried to find explicit rules of perspective that could be followed in a mechanical way. One of the first theoretical treatises that contained the correct rules for the construction of perspective was the booklet *Della Pittura* written by Leon Battista Alberti. Similar treatises were written during the fifteenth and sixteenth century by several painters, among others also by Albrecht Dürer. These treatises are interesting from the geometrical point of view, because we can find in them a remarkable linguistic innovation. When the painters tried to understand how perspective functions, they created a *representation of a representation*.

The use of representation of a representation is a characteristic feature of the new form of language, which I suggest calling the *projective form*. This form changed in a fundamental way the principles of the construction of a picture. The main achievement of the previous (perspectivist) form was a faithful representation of the theme of the painting from a particular point of view. Nevertheless, the point of view itself was not represented in the painting. It was the point out of which we had to look at the painting, and as such it remained unrepresented. The projective form brought a radical change with respect to the point of view: the point of view became explicitly represented in the painting. The *explicit representation of the point of view* brings the possibility of passing from one perspective to another. The projective form of language introduced explicit rules by means of which a situation that is represented in one perspective can be depicted from another. This change of perspective happens by means of a central projection and the center of this projection is the explicitly represented viewpoint. A third aspect of the projective form was the introduction of *double reference*, i.e., in a sense the replacement of the reality by its picture. From the epistemological point of view this was an important step that has many analogies in mathematics. For instance Frege in his *Foundations of Arithmetic* replaced the question *"What is a number?"* by the question *"When have two systems the same cardinality?"*. Thus Frege replaced the question about the reference of a numerical term by a question about the equivalence of two such expressions. It is interesting to notice that this idea appeared for the first time in geometry with Desargues (Frege actually acknowledged the geometric origin of this idea). The fourth aspect of the projective form of language was the introduction of *ideal objects*. In synthetic geometry, ideal objects have the form of infinitely remote points, in which parallels "meet". Their introduction was necessary in order to make the central projection a one-to-one mapping, and so to make it possible to compose two or more central projections.

The four linguistic innovations, which together marked the birth of the projective form of language enabled Desargues to create a first consistent theory of infinity. Thanks to the introduction of the *representation of a representation*, of the *point of view*, of the *double reference*, and of *ideal objects*, the language of geometry was enriched to such a degree that it became able to express things that were hitherto inexpressible. The notion of infinity that was considered incomprehensible and surpassing the capacities of the human mind became a subject of mathematical investigation. This change is important also from the point of view of the discovery of non-Euclidean geometries, because these geometries differ from the Euclidean one in their "behavior at infinity". Therefore the creation of a language capable of representing infinitely remote points was a step of paramount importance on the road to non-Euclidean systems.

2.1.2.1.   The Implicit Variant of the Projective Form – Dürer

Albrecht Dürer showed us in one of his drawings a method by which it is possible to create a perspectivist painting. I will describe Dürer's procedure in detail, because it enables me to show what is common and what is different in perspectivist and projective picturing. Imagine

that we want to paint some object so that its picture would evoke in the spectator exactly the same impression as if he were looking at the original object. Let us take a perfectly transparent foil, fix it onto a frame and put it between our eye and the object we intend to paint. We are going to dab paint onto the foil, point by point in the following way: We choose some point on the object (let it for instance be brown), mix paint of exactly the same color and dab it on that point of the foil, where the ray of light coming from the brown point of the object into our eye intersects the foil. If we have mixed the paint well, the dabbing of the paint onto the foil should not be visible. After some time spent by such dotting we create a picture of the object, which evokes exactly the same impression as the object itself. By a similar procedure the Renaissance painters discovered the principles of perspective. Among other things, they discovered that in order to evoke the illusion of two parallel lines, for instance two opposite sides of a ceiling, they had to draw two convergent lines. They discovered this but did not know why it was so. The answer to this, as well as many other questions, was given by projective geometry.

2.1.2.2.   The Explicit Variant of the Projective Form – Desargues

Gérard Desargues, the founder of projective geometry came up with an excellent idea. He *replaced the object with its picture*. So while the painters formulated the problem of perspective as a relation between the picture and reality, Desargues formulated it as a problem of the relation between two pictures. Suppose that we already have a perfect perspective picture of an object, for instance of a jug; and let us imagine a painter who wants to paint the jug using our dotting procedure. At a moment when he is not paying attention, we can replace the jug by its picture. If the picture is good, the painter should not notice it, and instead of painting a picture of a jug he could start to paint a picture of a picture of the jug. Exactly this was the starting point of projective geometry. The advantage brought by Desargues' idea is

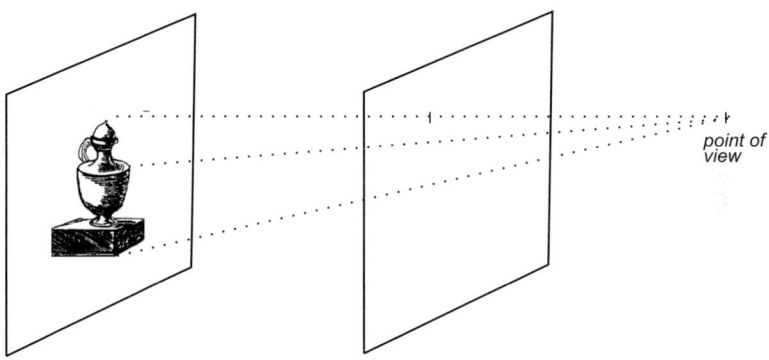

point of
view

that, instead of the relation between a three-dimensional object and its two-dimensional picture we have to deal with a relation between two two-dimensional pictures. After this replacement of the object by its picture, it is easy to see that our dotting procedure becomes a central projection of one picture onto the other with its center in our eye. I have mentioned all this only to make clear that the *center of projection represents the point of view from which the two pictures make the same impression.*

Before we start to consider the central projection of some geometrical objects, we have to clarify what happens with the whole plane on which these objects are drawn. It is not difficult to see that, with the exception of two parallel planes, the projection of a plane is not the whole plane. On the first plane (plane α – the plane from which we project) there is a line a of points for which there are no images. On the other hand, on the other plane (plane β – the plane onto which we project), there is a line b onto which nothing is projected. To make the central projection a mapping, Desargues had first of all to *supplement both planes with infinitely remote points.* After this the line a consists of those points of the plane α which are mapped onto the infinitely remote points of the second plane β. On the other hand, the line b

consists of images of the infinitely remote points of the plane α. So by supplementing each plane with infinitely remote points, the central projection becomes a one-to-one mapping. In this way Desargues created a technical tool for studying infinity. The idea is simple. The central

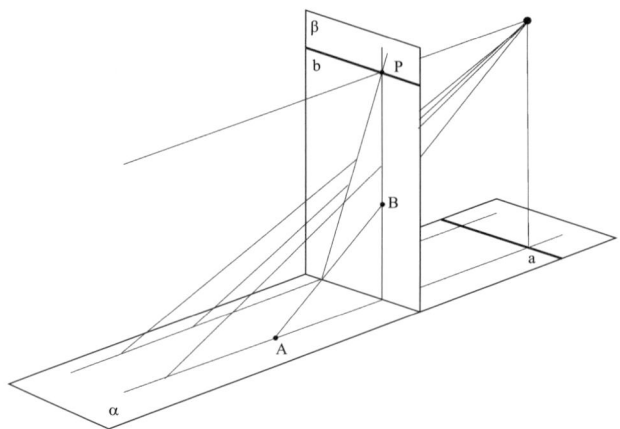

projection projects the infinitely remote points of the plane α onto the line b of the plane β. So, if we wish to investigate what happens at infinity with some object, we have to draw it on the plane α and project it onto β. If we draw two parallel lines on the plane α, we shall see that their images on the plane β intersect at one point of the line b. From this we can conclude that parallel lines also intersect on the plane α. They intersect at infinity, and the point of their intersection is mapped onto that point of the line b where their images intersect. If we draw a parabola on the plane α, we shall see that the parabola touches the infinitely remote line. This is the difference between the parabola and the hyperbola, which intersects the infinitely remote line at two points. So *Desargues found a way to give to the term infinity a clear, unambiguous and verifiable meaning.* Desargues' replacement of reality by its picture makes it possible to study transformations of the plane on which the objects are placed independently of the objects themselves. We could say, to study the transformations of the empty canvas. We have seen that exactly these rules of transformation of the plane enforce the fact that the images of parallel lines are not parallel. It is not an individual property of the lines themselves but the property of the plane on which these lines are placed. Euclidean geometry studied triangles, circles, etc., but these objects were, so to speak, situated in a void. In projective geometry the object becomes situated on the plane. Much of what happens to objects by the projection is determined by the rules of projection of the plane on which they are situated. We have seen this in the simplest case of the projection of two parallel lines. The point of intersection of their images is determined by the relation between the two planes α and β. So projective geometry investigates not only the objects themselves, but it also brings the background (the plane or the space) where these objects are situated, into the theory.

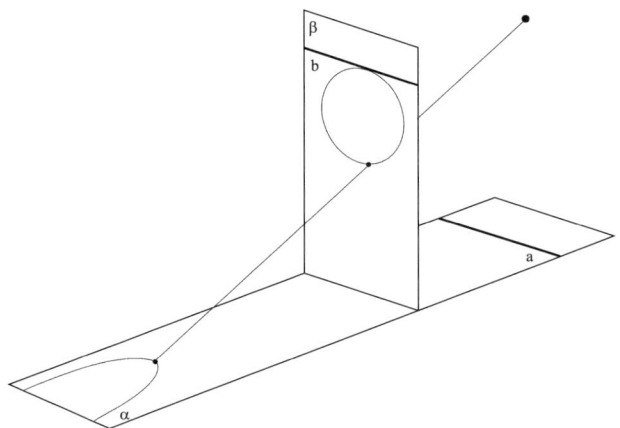

In pictures of projective geometry there is a remarkable point – different from all other points – the center of projection. As shown above, the center of projection represents in an abstract form the eye of the painter from Dürer's drawing. Besides this point, pictures of projective geometry contain also a remarkable straight line – the line b. The position of the line b on the plane β is determined by the center of projection that represents the eye of the spectator. It is not difficult to see that the line b represents the horizon. But it is important to realize one basic difference between the horizon in a perspectivist painting and in a picture of projective geometry. In projective geometry the horizon is a straight line, which means *it belongs to the language*. It is not something that shows itself only when the picture is completed, as in the case of paintings. Desargues drew the horizon, made from it an ordinary line, a sign of the iconic language.[4] There is nothing like

---

[4] The horizon is turned into a line already in Dürer. It became an expression of language, i.e., something that language can express and not only show. To see this, let us imagine that the painter depicted in Dürer's picture is painting, instead of a painting of a jug, a painting of a landscape. Then on his foil gradually a horizon would appear. For the painter represented by Dürer this horizon would be

the center of projection or the horizon in Euclidean geometry. A Euclidean plane is homogeneous, all its lines are equivalent. So instead of *the Euclidean looking from nowhere onto a homogeneous world, or the perspectivist watching from outside, for Desargues the point of view is explicitly incorporated into language.* It is present in the form of the center of projection and of the horizon which belongs to this center.

This incorporation of the point of view into the theory made it possible for Desargues to broaden qualitatively the concept of geometrical transformation. We cannot say that Euclid did not use transformations. In some constructions he used rotations or translations. But since he did not have the point of view incorporated into the language of his theory, he could define only a few transformations. To define a transformation means to specify what changes and what remains unchanged. As the point of view was not contained in the language of his theory, Euclid had to define his transformations uniformly with respect to any point of view. This means that his transformations could not change the shape of the geometrical object. Desargues, having explicitly introduced the point of view in his theory, was able to define a much larger class of transformations. He could define what is changed and what remained unchanged with respect to a unique point. Exactly thus a projective transformation is defined: two figures are projectively equivalent if there is a point from which they appear the same.

2.1.3. The Interpretative Form of Language of Synthetic Geometry

It is an interesting historical fact that even though Saccheri and Lambert discovered many propositions of non-Euclidean geometry, they

---

only shown by his painting and therefore it would belong to its form. But because Dürer's picture is a representation of a representation and thus the process of painting of the landscape is represented from an external viewpoint, for this external viewpoint the horizon on the painting does not belong to the form. It is turned into an ordinary line which is simply represented in the external perspective. Thus the form of language of the painting represented in the picture is objectified.

This objectification of the horizon is even stronger in Desargues, who replaced reality by its picture. In Desargues the painter paints a painting of a picture and so already for the painter himself, and not just for the external spectator, the horizon is objectified. To decide whether the horizon is turned into an object or belongs to the form, it is sufficient to realize that the real horizon moves inside the country when we move our viewpoint. In contrast to this a horizon depicted on the picture by which Desargues replaced the landscape when he analyzed the painting of the picture of a picture cannot move in the picture. The horizon is replaced by a fixed line, i.e., it is turned into an object.

persisted in believing that the only possible geometry is the Euclidean one. The breakthrough in this question started only with Gauss, Bolyai and Lobachevski, who in the first half of the nineteenth century came to the conviction that besides the Euclidean geometry also another geometry is possible. Gauss first called the new geometry anti-Euclidean, then astral, and later invented the name non-Euclidean, which is currently used. The most striking point about this geometry is the fact that many of its theorems, together with their proofs, were known to Saccheri and Lambert. So it could seem that the contribution of Gauss, Bolyai and Lobachevski was not a mathematical one but more a psychological one, consisting in a change of attitude towards the new geometry. While Saccheri and Lambert rejected it, Gauss, Bolyai and Lobachevski accepted it.

But a psychological explanation is unsatisfactory. Our task is to ask what made this change of attitude possible. We have to find a reconstruction of this discovery in terms of the picture theory of meaning. In the case of Lobachevski it is possible to offer such a reconstruction, based on his derivation of the trigonometric formulas of non-Euclidean geometry. This derivation played an important role for Lobachevski, because if it is possible to do trigonometric calculations in non-Euclidean geometry, nothing could go wrong. But more importantly, this helps us to translate a psychological fact of attitude change into an epistemological fact of scientific discovery.[5]

2.1.3.1. *The Implicit Variant of the Interpretative Form – Lobachevski*

In non-Euclidean geometry there are two objects, which are similar to straight lines. One is the so-called *equidistant line*, which is the set of all points lying on one side of a straight line at a constant distance from it. The second is the so-called *limit line*, which is the set of points towards which a circle passing through a fixed point A approaches when its diameter grows beyond any boundary. In Euclidean geometry both the equidistant line and the limit line are straight lines. In non-Euclidean geometry, on the other hand, they are not straight lines. Many of the incorrect proofs of Euclid's fifth postulate were based on the mistake that the geometers constructed an equidistant line but regarded it as a straight line. But to say that the equidistant line is a straight line means to accept Euclid's fifth postulate. These proposi-

---

[5] A concise exposition of the geometry of Lobachevski can be found in (Shirokov 1955).

tions are equivalent. Therefore every proof of Euclid's fifth postulate based on the assertion that the equidistant line is a straight line is a circular proof.

In a similar way one dimension higher, we can form an *equidistant surface* to a plane and a *limit surface* (from a sphere passing through a fixed point *A*, whose diameter grows beyond any boundary). On the limit surface there are limit lines lying in a similar way to straight lines lying on the plane. Lobachevski discovered an interesting circumstance, namely, that on the limit surface an analog to Euclid's fifth postulate holds. If we choose a limit line *l* and a point *P* not lying on this line of the limit surface, then there is exactly one limit line, passing through the point *P* not intersecting the line *l*. That means that on the limit surface, for limit lines taken instead of ordinary lines, the whole of Euclidean geometry holds. So does among others the cosine theorem:

$$a^2 = b^2 + c^2 - 2bc \cdot \cos \alpha \qquad (2.1)$$

where a, b, c are the lengths of the sections of the limit lines forming the sides of the triangle and α is the magnitude of the angle by the vertex *A*. The formula (2.1) holds on the limit surface. Lobachevski

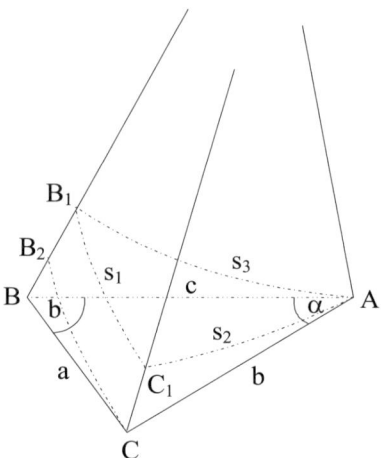

wanted to derive the trigonometric formulas for triangles whose sides are sections of the straight lines of the non-Euclidean plane and not segments of limit lines.[6] For this purpose he used the above picture (Lobachevski 1829, p. 112). In it ABC is a triangle on the non-Euclidean plane, which touches the limit surface in the point A. AB_1C_1 is the projection of the triangle ABC onto the limit surface. For the triangle AB_1C_1, since it is a triangle on the limit surface, Euclidean geometry holds. The cosine theorem for this triangle gets the form:

$$s_1^2 = s_2^2 + s_3^2 - 2s_2s_3 \cos \alpha \qquad (2.2)$$

The problem now is how to transfer this formula from the limit surface onto the non-Euclidean plane. Lobachevski succeeded in finding the formulas that connect the lengths of the straight line segments of the plane with the lengths of their projections on the limit surface. These formulas are:

$$s_3 = \sigma \cdot \tanh \frac{c}{k}, \quad s_2 = \sigma \tanh \frac{b}{k}, \quad s_1 = \sigma \, \frac{\tanh \frac{a}{k}}{\cosh \frac{b}{k}}. \qquad (2.3)$$

In these formulas there are two constants k and σ. The constant k is called the radius of curvature of the non-Euclidean plane, and it plays an analogous role to the radius of the sphere in the formulas of spherical geometry. The constant 2σ is the length of the segment of the limit line, onto which the whole straight line of the plane is projected by this projection. The functions $\sinh(x)$, $\cosh(x)$, and $\tanh(x)$ are the so-called hyperbolic sine, hyperbolic cosine and the hyperbolic tangent. Using formulas (2.3) it is possible to transfer the cosine theorem (2.2) from the limit surface onto the non-Euclidean plane. It is sufficient to substitute into the formula (2.2) for the s_1, s_2, and s_3 their values given by (2.3). After some elementary transformations we obtain:

$$\cosh \frac{a}{k} = \cosh \frac{b}{k} \cdot \cosh \frac{c}{k} - \sinh \frac{b}{k} \cdot \sinh \frac{c}{k} \cdot \cos \alpha. \qquad (2.4)$$

This formula is remarkable because if we expand the hyperbolic functions into infinite series,[7] and let $k \to \infty$, we obtain from the non-

---

[6] The relation (2.1) is not related to the picture. It is a general formulation of the cosine theorem in the usual notation. The variables a, b, and c in the picture do not satisfy the relation (2.1) because this relation belongs to Euclidean geometry while a, b, and c are the sides of a non-Euclidean triangle. Thus the formula (2.1) should be kept separate from the following derivation.

[7] Series for the hyperbolic functions have the following form:

$$\cosh\left(\frac{b}{k}\right) = 1 + \frac{b^2}{2k^2} + \frac{b^4}{24k^4} + \dots, \quad \sinh\left(\frac{b}{k}\right) = \frac{b}{k} + \frac{b^3}{6k^3} + \dots$$

The derivation of them can be found for instance in Courant 1927.

Euclidean formula (2.4) the Euclidean formula (2.1). That means that Euclidean geometry is the limiting case of the non-Euclidean one for the radius of curvature of the non-Euclidean plane growing to infinity. Thus the formula (2.4) indicates a deeper coherence between the two geometries. Besides this the formula (2.4) connects magnitudes of the angles with lengths of the segments. Therefore it plays a central role in all of trigonometry.

But how did Lobachevski derive this formula? First he *embedded* a fragment of Euclidean geometry (in the form of the limit surface) into non-Euclidean space, and then he *transmitted* the geometrical relations from this fragment onto the non-Euclidean plane. So this picture represents a junction of two languages. These two languages are separated; they appear on different backgrounds. The Euclidean language is situated on the limit surface; the non-Euclidean one is situated on the plane. The formulas (2.3) establish the translation between these two languages. I think that the structure of this picture goes beyond the possibilities of the language of projective geometry. In Desargues both the object and its image in the projection are placed on equal, Euclidean, backgrounds. So the whole projection takes place in one language, the Euclidean. In Lobachevski, on the contrary, the objects are non-Euclidean (the triangle ABC) and the images are Euclidean (the triangle AB_1C_1). So the projection is a translation between two different languages. The formulas (2.3) are neither formulas of Euclidean geometry, nor formulas of non-Euclidean geometry. They belong to the metatheory, connecting these two geometries. This is something qualitatively new.

But besides this change of background there is another, maybe even more striking change brought about by Lobachevski. *How was it possible to draw these pictures?* The line AC_1, the one which is drawn on the picture, cannot be a segment of any limit line. It is drawn on ordinary Euclidean "paper", and there does not exist any limit line on this paper. In order to be able to understand this picture, we have to know that we must not take what we are looking at literally. Sure, we are looking at an ordinary network of lines of the Euclidean plane, and on the Euclidean plane there are no limit lines. But the situation in many respects resembles that of Renaissance paintings. A painting, strictly speaking, does not have any depth as well, but nevertheless it is able to evoke depth. By interpretation of perspectivist painting, we have explained this ability to evoke depth as the manifestation of a new form of language and of the epistemic subject related to this new form. This subject had the form of the point of view, and from this point of view we see – looking at two straight lines which obviously converge and intersect – two parallel sides of a ceiling.

It seems that also in the case of Lobachevski we are dealing with a new form of language. The introduction of this new form of language consists in taking up of an *interpretative distance*. This distance enables us to see a non-Euclidean triangle – which it is impossible to draw – behind the triangle ABC. The triangle ABC itself is, in the picture, present in the form of an ordinary Euclidean triangle. The interpretative distance becomes part of the language in the sense that in the drawing of the picture it is taken into consideration. Anyone unable to take up this interpretative standpoint cannot understand the drawings of Lobachevski. The new form of language that shows itself in the figures of the non-Euclidean geometry I suggest calling the *interpretative form*. The basic problem of the pictures of Lobachevski is that they contain a conflict: the language that they use is Euclidean, but what they want to express is non-Euclidean.

2.1.3.2. *The Explicit Variant of the Interpretative Form – Beltrami*

The consistency of non-Euclidean geometry was proven in 1868, twelve years after Lobachevski's death, by Beltrami, who constructed its first model. A simplified version of this model was suggested by Klein in 1871, and I would like to discuss it briefly. But before we turn to Beltrami's model, we must return to the picture which was used in the derivation of the trigonometric formulas by Lobachevski. There we drew only a small sector of the limit surface in the form of the triangle AB_1C_1. Let us now draw a greater part of this surface as well as of the plane. We see that the plane is projected not onto the whole

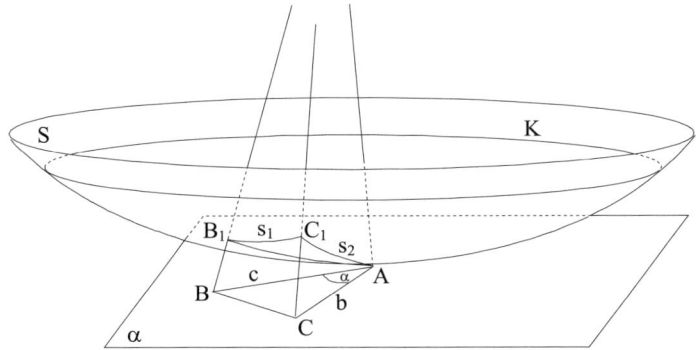

limit surface, but only onto part of it bounded by the circle K. The parameter σ in the formulas (2.3) is the diameter of this circle. The limit surface is a sphere of non-Euclidean geometry, whose center is infinitely remote. Lobachevski's projection from the limit surface onto the plane happens from this infinitely remote center. In our discussion of Desargues we have mentioned that the center of projection represents the point of view. The problem with the above figure is that it tries to express a non-Euclidean content using Euclidean means. Beltrami removed this problem when he looked at the whole picture from its center of projection. What did he see? We know that the center of

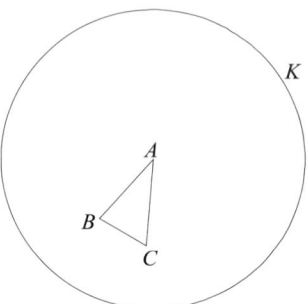

projection is that point from which the image and the original make exactly the same impression. That means that from this point the limit surface (more exactly those parts of it onto which the whole plane is projected) and the non-Euclidean plane look exactly the same. The triangles ABC and AB_1C_1 blend. The Beltrami–Klein model is based on an original idea, which resembles Desargues' idea of replacing the reality by its picture. Only instead of replacing reality by a picture, in the Beltrami–Klein model we *stick the picture onto the original*. It could seem that, by this identification of the picture and the original, the information gets completely lost. Lobachevski's transition of the trigonometric formulas from the limit surface onto the plane was based on the fact that these objects were different, and so they could correspond to one another.

But here Beltrami had recourse to the *interpretative subject*. The difference between the original and the picture, which he identified physically, was transferred onto the shoulders of the interpretative subject. In a manner similar to Desargues when he turned the implicit perspectivist subject of the Renaissance paintings into the explicit subject of projective geometry having the form of the center of projection, Beltrami *objectified the interpretative subject* of Lobachevski. He turned the implicit requirements of Lobachevski, which required one to see behind Euclidean lines non-Euclidean objects, into explicit rules. So the interpretation becomes expressible within the language. In both cases something implicit, belonging only to the understanding of language, became expressible explicitly in the language. In this way the subject becomes part of the language. We see that the changes in geometry that we analyze in this chapter consist in turning deeper and deeper layers of our subjectivity into structures explicitly present in language. First these layers are uniformly connected to the language in the form of an *implicit* (perspectivist, projective, or interpretative) subject that brings some new way of understanding the pictures. Then this implicit subject becomes *explicitly expressed* in the language and so the corresponding layer of our subjectivity is turned into an object (like the center of projection). I suggest calling these changes *relativizations*. In their course, deeper and deeper layers of our subjectivity are objectified. They are represented in the language by means of explicit expressions, which are used as if they were denoting some real objects.

But let us come back to the Beltrami–Klein model. When we look at the picture used by Lobachevski in his derivation of the trigonometric formulas from the improper center of projection, the Euclidean objects on the limit surface blend with the non-Euclidean objects of the plane (for instance the triangles ABC and AB_1C_1 appear exactly the same). *But that means that it is possible to draw them!* The conflict of Lobachevski's drawings is overcome. We can draw everything "in a Euclidean way" and interpret it "in non-Euclidean terms". This is the advantage of the fact that the interpretative subject is directly present within the language. The interpretation, which for Lobachevski had the form of implicit understanding of what the picture had to express, is turned here into explicit naming. I can draw a Euclidean object and interpret it *by definition* as the non-Euclidean object with which it is identified in the projection. So this explicit incorporation of the interpretation in the language makes it possible to represent the non-Euclidean plane. We in fact draw the circle of the limit surface which, as Lobachevski discovered, is Euclidean. The non-Euclidean plane is projected onto this circle, and thus from the center of projection they make exactly the same impression. Thus we name the geometrical objects inside of this circle after those non-Euclidean objects with which these Euclidean – and therefore drawable – objects blend. The naming, which is normally something implicit, based on showing, happens here explicitly, in the language.

The Beltrami–Klein model is thus based on a simple picture, in which the inside of a circle with some of its chords is represented. This picture is complemented by explicit rules of interpretation:

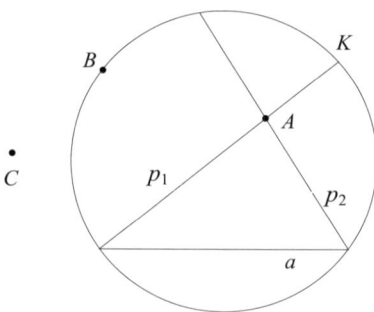

	External Language	Internal Language
K	a circle on the Euclidean plane	the horizon of the non-Euclidean plane
A	a point inside the circle	a point on the plane
B	a point on the circular line	infinitely remote point
C	a point outside the circle	????????
a	a chord of the circle	an infinite straight line
p_1	a chord not intersecting the chord a	a parallel to the straight line a

The points inside the circle (the circle excluding the circular line) represent the whole non-Euclidean plane, and the chords (excluding their endpoints) represent its straight lines. In this model all the axioms of Euclidean geometry are satisfied. Through any two different points of this plane there passes exactly one straight line, etc. – all axioms with the exception of the axiom of parallels. If we choose any straight line, for instance a, and any point which does not lie on this line, for instance A, through this point we can thus draw two straight lines (p_1 and p_2) which do not intersect the line a.

Even though we are not able to see the non-Euclidean world, the model can represent it. We are unable to depict the non-Euclidean world, we can only model it. The model is a qualitatively new way of representation. It is new in comparison to Lobachevski's, because there is no conflict in it. It is also new compared with Desargues'. In Desargues' each object was situated on one background. Beltrami, by sticking the picture onto the original, gives to each object two backgrounds. In his model Beltrami takes a Euclidean plane; and on this plane he first represents the non-Euclidean background in the form of a circle, which represents the horizon of a non-Euclidean plane. Then on this non-Euclidean background he represents the objects. So each object gets two frames; it is situated on two backgrounds simultaneously. On the one hand, it is situated on the external background of the Euclidean plane, on which it is a chord of a circle. But on the other hand the same object is situated also on the internal background, where it acts as a straight line of the non-Euclidean plane.

This double co-ordination of objects made it possible to prove the consistency of non-Euclidean geometry. If non-Euclidean geometry was inconsistent, it would contain some theorem about the points and straight lines of the non-Euclidean plane that could be proven together with its negation. Beltrami's model makes it possible to translate this non-Euclidean theorem into the language of Euclidean geometry that would state the same things, not about points and lines of the non-Euclidean plane, but about the points and chords of the Euclidean circle K. It is not difficult to see that this translation transforms proofs into proofs, and so we would get a theorem of Euclidean geometry, that also could be proven together with its negation. So if non-Euclidean geometry was inconsistent, Euclidean would be as well.

2.1.4. The Integrative Form of Language of Synthetic Geometry

Although Beltrami's model definitely removed all doubts concerning the consistency of non-Euclidean geometry, nevertheless this is a model of non-Euclidean geometry *within* Euclidean geometry. This means that these geometries do not have equal status. On the contrary, Euclidean geometry is the presumption of the possibility of the non-Euclidean geometry. First a Euclidean plane must be given, in order to draw on it the circle K and to make a model of the non-Euclidean plane inside this circle. This means that Euclidean geometry is the transcendental presumption, in the sense of Kant, of non-Euclidean geometry. It is obvious, that this aspect of the Beltrami–Klein model renders a wide space for the defense of Kant's philosophy of geometry, and this must have caused delight to every neo-Kantian. It is very probable, that if Kant would have lived long enough to have learned about the existence of non-Euclidean geometry, he would have been easily able to adapt his system to this fact. Indeed, in the Beltramian model Euclidean geometry is the transcendental presumption of the non-Euclidean, and so

its a-priori character is not violated. Thus Euclidean geometry can preserve its status of an a-priori form of spatial intuition. We have only to
add to this function, formulated by Kant, a new one. Euclidean geometry is also the a-priori form of constructions of non-Euclidean models.
On the other hand it is also obvious that this inequality of the status of
Euclidean and non-Euclidean geometries in the Beltramian model has
its origin more in our physical constitution than in the nature of these
geometries. So the primacy of the Euclidean geometry, which lies at
the basis of Beltrami's model, is more a weakness than a strength of
this model.

2.1.4.1.   *The Implicit Variant of the Integrative Form – Cayley*

The English mathematician Cayley had the idea to look at Euclidean
geometry from the non-Euclidean position, that is, to reverse the order in which they enter into the construction of Beltrami's model.
He wanted to stop seeing Euclidean geometry as something given, as
something binding, and try to see it also as a model. But how can we
do this? Cayley's basic idea emerged when he examined the problem
of how to introduce the concept of distance into Beltrami's model. Let
us imagine small beings, for which the interior of the circle K is their
whole world. How would they measure distances? Our usual metric
is not suitable, because in it the distance from a point inside the circle K to its circumference is finite, but for our beings this distance
should be infinite, because the circumference of the circle is the horizon of their world, and in any world the horizon is infinitely remote.
So the usual concept of distance is unsatisfactory. Cayley suggested

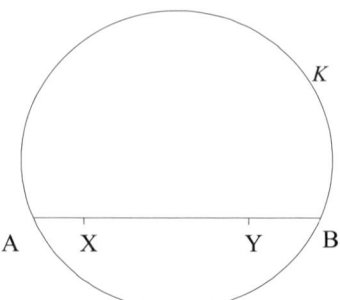

looking at this circle from the standpoint of projective geometry. The
points A and B, in which the line XY intersects the circumference
of the circle K, do not belong to the world of our beings (for them

these points are infinitely remote), but for us they are readily available. Here the interpretation leaves the internal language of the model and resorts to its external language. If we are looking for the concept of distance between points X and Y, we have four points available (in the external language), namely X, Y, A and B. Four points define a projective invariant magnitude, which was already discovered by Desargues, namely the cross-ratio:

$$(A, B; X, Y) = \frac{|AX|}{|BX|} : \frac{|AY|}{|BY|}.$$

For the sake of brevity I omit the historical details and present only the resultant formula, which expresses the distance between points X and Y in the model as:

$$d(X, Y) = |\ln(A, B; X, Y)|.$$

It is not difficult to see that when the point X approaches the point A, the cross-ratio converges to zero, its logarithm drops to minus infinity, and so in absolute magnitude we get the value we need. As the point X approaches the horizon, its distance from the point Y grows beyond any bound. Of course, this formula is not in the language of our beings. They do not understand what the point A is. But this is not important. The whole of Beltrami's model is formulated in the external language. What is important is that the distance is formulated with the use of a projective invariant, namely the cross-ratio.

The suggestion of Cayley was to *look at the whole of Beltrami's model and not only at its metric from the standpoint of projective geometry.* Let us forget that Beltrami's model is constructed on a *Euclidean* plane and imagine that it is constructed on a *projective* plane. What is a projective plane? It is what Desargues made from the Euclidean plane. So let us forget the parallels, the distances and the angles. What remains is a snarl of lines which intersect each other. Now onto this "clean" plane we draw the circle K. This circle intersects every straight line at two points, and with the help of these points we introduce the concept of distance, just as above. So the circle introduces a non-Euclidean metric. Besides this the circle divides all pairs of straight lines into three groups: those which intersect inside the circle, those which intersect onto its circumference, and those which intersect outside the circle. Those of first kind are secants, the second are parallels, and the third are a special kind of lines, which Lobachevski called *divergents*. So we see that besides the concept of distance, the circle can also be used to introduce the concept of parallels and the other kinds of relationships between straight lines. Also the concept of angles can be introduced using the circle K.

So the circle K is that object which induces the non-Euclidean structure onto the projective plane. Instead of Beltrami's construction

$$\mathbf{E} \rightarrow \mathbf{L}$$

in which the model of the non-Euclidean geometry is based on the Euclidean plane, Cayley suggested a different scheme

$$\mathbf{E} \rightarrow \mathbf{P} \rightarrow \mathbf{L}.$$

The first arrow indicates the transition from a Euclidean plane to a projective plane and it consists in *disregarding* its Euclidean structure, i.e., disregarding the Euclidean concepts of parallels, distances and magnitudes of angles. The second arrow indicates the shift from a projective plane to a non-Euclidean plane, and it consists in the *introduction* of the non-Euclidean structure, that is, introducing the non-Euclidean concepts of parallels, distances and magnitudes of angles. This second step is based on the circle K, which Cayley has called the *absolute*. Cayley comprehended the role that the circle K played in Beltrami's model. It constituted the non-Euclidean structure of the model's geometry. So he was able to formulate a fundamentally new question, namely the question: What constitutes the Euclidean-ness of the Euclidean plane? In the framework of Beltrami's model it was not possible to ask this question. If we take the Euclidean plane as something given, then the question of how we can introduce Euclidean geometry onto this plane is meaningless. The transition

$$\mathbf{E} \rightarrow \mathbf{E}$$

is impossible to describe. What constitutes a language is inexpressible in this language. In Cayley's framework the same question arises very naturally, namely what constitutes the Euclidean-ness of the Euclidean plane. It asks what object should replace Beltrami's circle K in order to get the scheme

$$\mathbf{E} \rightarrow \mathbf{P} \rightarrow \mathbf{E},$$

that is, in order to get Euclidean geometry again in the projective plane. The answer to this question is striking. The absolute of a Euclidean plane is degenerate: it consists of two imaginary points (i.e., points with complex co-ordinates) lying on a real line (i.e., a line whose equation has real coefficients). So what constitutes our world are two imaginary points. Isn't that strange? In addition let us remark that, in the theory of relativity, the time co-ordinate is also imaginary. Is this merely chance? And most of all, the absolute does not determine Euclidean geometry completely. It is necessary to add additional conditions.

But let us stop speculating and return to epistemology. We have seen that Cayley brought a qualitatively deeper insight into the structure of Euclidean geometry. He was able to find what constitutes its Euclidean structure. But what made this fundamentally new insight possible? We have seen that it was a seemingly small shift, namely the transition from the Euclidean to the projective plane as the basis of the model. But how did Cayley accomplish this transition? By an *appeal!* He did not change anything in Beltrami's picture, rather he just asked us to forget that it is drawn on the Euclidean plane and instead to see the whole picture as drawn on the projective plane. That means we must forget the parallels, the distances, and the magnitudes of angles. But who can do this? Nobody, I think. It seems that we cannot see except by using the Euclidean framework.

So here we are dealing with an appeal, similar to that on which the whole of perspectivist painting is based – in that case, to see two parallel sides of a ceiling behind the two intersecting lines of the painting. The appeal of Lobachevski was analogous, asking us to see behind the lines of a Euclidean plane the objects of non-Euclidean geometry. In a way similar to Dürer or Lobachevski, Cayley is also asking us to abandon what we are looking at, namely a Euclidean plane, and see instead the projective plane. Thus we can interpret this appeal just as we did in the other two cases, namely as the introduction of a new implicit form language.[8] It is implicit, because the transition from the Euclidean plane to the projective plane is not explicitly described; Cayley does not tell us what we have to do in order to see the projective structure. He only requires an implicit understanding of what would happen if we were able to see the plane without parallels, distances and magnitudes

---

[8] I would like to call a universal structure, which in the case of perspectivist painting and projective geometry had the form of a viewpoint, an epistemic subject. In Lobachevski or Cayley it is not correct to use the term viewpoint because they are representing non-Euclidean systems that cannot be viewed from any particular viewpoint. Nevertheless, in these abstract systems there is still some subjective pole to which the language is related. This pole does not have the form of a point from which we have to look at the representation. It has rather the form of a particular interpretative stance or distance which we have to take. Therefore it seems more suitable to call the subjective pole of a language not viewpoint but *epistemic subject*.

of angles. Of course, strictly speaking, we are unable to see this way, but we understand what he wants to tell us. If we are willing to follow him, we are then able to understand the fundamental question of what constitutes the Euclidean structure of Euclidean geometry, just as depth was revealed within Renaissance painting and just as we got an insight into the non-Euclidean world through the pictures of Lobachevski. In this way Euclidean geometry stops being something given, something a-priori, and we can question its fundamental nature.

I would like to call the implicit form of language introduced by Cayley the *integrative form*, because it makes it possible to integrate Euclidean and non-Euclidean geometry into one system. From this point of view we see the common foundation of the projective plane, from which both Euclidean and non-Euclidean geometries emerge in a uniform way, with the help of the curve called the absolute by Cayley. In this framework Euclidean geometry is no longer a presumption of the possibility of non-Euclidean geometry. They both are equal; they both arise in the same way. The fact that Euclidean geometry had a prior position in our world is only a question of physics and psychology, but not of geometry. From Cayley on, Euclidean and non-Euclidean geometries are on a par.

Cayley's transition to the projective plane as the base of geometry makes it possible to go further to an even more radical question: *What geometries are possible?*

$$\mathbf{E} \to \mathbf{P} \to ?$$

This is a matter of taking different curves in the role of the absolute, and finding out what kind of geometry they induce on the projective plane. This is a fundamentally new question, namely the question: What is geometry? Cayley formulated this question, but the answer to it was given by Klein.

2.1.4.2.   *The Explicit Variant of the Integrative Form – Klein*

Let us consider an analogous situation from mathematical logic. There the question arose, which other binary logical connectives besides implication, conjunction and disjunction are possible. This question could be answered when the logical connectives were *identified* with Boolean functions. It is not difficult to see that there are 16 such connectives.

It is easy to recognize the second function as disjunction, and the fourth as implication.

p	q	1	2	3	4	5	6	7	8	9	10	11	12	13	14	15	16
1	1	1	1	1	1	0	1	1	1	0	0	0	1	0	0	0	0
1	0	1	1	1	0	1	1	0	0	1	1	0	0	1	0	0	0
0	1	1	1	0	1	1	0	1	0	1	0	1	0	0	1	0	0
0	0	1	0	1	1	1	0	0	1	0	1	0	1	0	0	1	0

The analogical question about geometries was formulated by Cayley. *Which geometries are possible?* And Klein found the identification which made it possible to answer this question. In Cayley's framework the question meant: Which curves taken as the absolute give rise to a geometry? Cayley understood that the rule of the circle in Beltrami's model constitutes the concept of distance, and so he thought that all possible geometries could be found by examining different curves. But we have seen that this approach is unsuitable for analyzing even Euclidean geometry. The absolute is in this case degenerate and for this reason it does not create any metric. So, for the degenerate cases, and the most interesting cases are degenerate, we need another approach. In addition, there are too many curves, many more than there could be geometries.

Klein found a way out. The circle K in Beltrami's model has an additional property which Cayley did not notice. In the group of all projective transformations, the circle defines a subgroup of those transformations which transform the circle onto itself. By identifying geometries with the subgroups of the projective group, Klein found a tool which made it possible to give an answer to Cayley's question. Not every curve is suitable for an absolute. The absolute can only be a curve such that it defines a subgroup within the projective group. Klein thus replaced Cayley's implicit scheme

$$\mathbf{E} \to \mathbf{P} \to \mathbf{L}$$

by the explicit one

$$\mathbf{G_E} \to \mathbf{G_P} \to \mathbf{G_L} .$$

Thus Klein did the same thing with Cayley's implicit integrative form of language that Beltrami had done with Lobachevski's implicit interpretative form of language, and Desargues with the implicit perspectivist form of language of Dürer: he *incorporated it into the language*. The transformation group is exactly the tool that makes it possible to replace the appeal of Cayley to "forget the parallels, distances and angles", by an explicit instruction, "from the Euclidean group transit to the projective one". That is because the projective group is exactly that group which "destroys" the parallels, the distances and the angles, leaving only the intersections and the cross-ratio.

2.1.4.3.   *Philosophical Reflection of the Integrative Form – Poincaré*

Let us now come to a short epistemological analysis of Klein's *Erlanger program* (Klein 1872), which opened a unifying view onto the whole of geometry using methods of group theory. Our task is to explain why exactly the theory of groups was so successful in unifying geometry. At first sight Klein's approach seems to be that he has taken a concept from algebra and brought it into geometry. So, why exactly this concept? Could he have taken another? Would it lead to a different development of geometry? What we need is an epistemological interpretation of the concept of group itself. The key to this lies in understanding why it was necessary to wait for the theory of groups until the nineteenth century. The answer to this question we already know. In Greek mathematics the concept of transformation was too narrow and it was only in projective geometry that this concept became broad enough to make it possible to study the transformations themselves. But what made it possible for Desargues to change the concept of transformation so radically? It was the incorporation of the point of view into the language of geometry. Somewhere here we should search for the starting point of an epistemological reconstruction of the concept of group. And really, the reconstruction we need can be found in the book *La Science et l'Hypothèse* (Poincaré 1902, p. 71).

In his book, Poincaré investigates the relation between geometrical space and the space of our sensory perceptions. At first sight it could seem that these two spaces are identical. But this is not the case. Geometrical space is continuous, infinite, homogeneous, isotropic and three-dimensional. Poincaré shows that the space of our visual perceptions is neither homogeneous, nor isotropic nor three-dimensional. That means that we cannot derive the concept of geometrical space from our visual perceptions alone. With the other sensory perceptions the situation is similar. That means that we cannot derive the concept of geometrical space from the space of any one isolated sensory organ. It is important to realize that, for the birth of geometrical vision, visual impressions are not enough. The tactile and motor perceptions are necessary as well. So from the very beginning the concept of geometrical space is connected with our body; and so, trying to derive it from the visual perceptions alone, as often done in phenomenology or empiricism, is an epistemological error.

Poincaré asserts that we have derived the concept of geometrical space from the relations in which the changes of different kinds of perceptions (visual, tactile and motor) follow each other. The most important among these relations are the relations of *compensation*. We can compensate changes in the visual field by motion of our body or eyes. Poincaré showed that these compensations have the structure of a group, and that this group is the group of transformations of Euclidean space. So Euclidean space is neither the space of our sight, nor the space of our touch, nor the space in which we are moving. The Euclidean space is the structure in which these three sensorial spaces are integrated together. This analysis shows that the concept of group is something deeply concerned with us. The Euclidean group is the tool with the help of which each of us transcends the private world of his or her sensory perceptions. In this way the concept of a group forms the ground on which the inter-subjective language of spatial relations is based. That means that the theory of groups, which is the theory making possible the transition from a series of impressions of isolated sensory organs to the invariants of the structure of their compensatory relations, is in a fundamental way present not only in the concept of space, but also in the concept of the object, reality, ...

Given this analysis, it is not surprising that Klein could use the concept of group in geometry, and that, with the help of the theory of groups, geometry reached a qualitatively higher level of abstraction. It is not surprising because the concept of group forms the foundation on which the concept of space is constituted. So Klein did not introduce an algebraic concept into geometry, rather he only made this concept explicit, which from the very beginnings lies implicit within the foundations of geometry. So we can say that Klein's *Erlanger program* was successful in bringing a unifying view to geometry, because the concept of group, on which this program is based, forms the epistemological foundation of the concept of space.

Now let us return to Poincaré's analysis of the concept of space. Poincaré speaks about compensatory relations between perceptions. But what does it mean to compensate? To compensate means to attain a sameness between the original and the new perceptions. But that is familiar to us, isn't it? Let us recall Dürer. He also wanted two perceptions to be the same, and he introduced the point of view as that point from which the picture and the original make exactly the same impression. So also in the case of compensation we are dealing with some "point of view", a "point of view" from which the perceptions before and after the compensation are the same. But this "point of view", which constitutes the group of compensatory relations and which forms the basis of Klein's approach to geometry, does not have the form of a point, as it did in the case of Desargues. For Desargues it was enough to introduce a single point into geometrical language. Nor has it the form of two viewpoints, the external and the internal, as it did in the case of Beltrami. For Beltrami it was enough to add a dictionary, which made it possible to translate theorems from external into internal language and vice versa. In Klein's case viewpoints fill the whole space. *Euclidean space can be seen as the space of viewpoints. It is not the space of the seen. The Euclidean space is not the space where the things at which we are looking are situated. It is the space of seeing. It is the space of the possible viewpoints from which we are looking at the world.* The transformation group integrates these viewpoints into one whole, which together with the group represent the epistemic subject of the language.

In his characterization of a geometry (Euclidean, non-Euclidean, or any other) Klein took a particular transformation group as the starting point. He defined the geometry corresponding to the group as the system of all propositions for which the truth value is not altered when we apply to the objects any transformation belonging to the group. In other words, the geometry that corresponds to a given group G is the study of those properties that are invariant with respect to the transformations belonging to the group. Thus Klein passed from the *interpretative form* of language, which guaranteed the translation from the internal language of the model to the external language (i.e. it guaranteed the equivalence of meaning between the corresponding expressions of these two languages), to the *integrative form*. The integrative form of language unites all possible geometries – the Euclidean, projective, affine, non-Euclidean, etc. – into one whole. Klein thus introduced a unity into geometry.

The Euclidean and non-Euclidean geometries are in contradiction when we interpret them as alternative descriptions of reality. If one of them is true, the other one must be false. Beltrami weakened this tension when he interpreted the different geometries as alternative *models* of reality. Each model is consistent; and the different models can be interpreted by means of the others (i.e. there exists a translation between them). Thus, even if only one of the different geometries can be true, the other geometries need not be discarded. They must be tolerated as consistent alternative systems. Klein made a further important step in understanding of the unity of geometry when he showed, that these different alternative geometrical systems complement each other. Thus the assertions of the particular geometries *contradict each other*; their models *tolerate each other*; and their transformation groups *complement each other*. The higher the level of abstraction of our form of language, the stronger the unity of the view it offers. The transformations of particular geometries are all elements of one universal group, the group of projective transformations. There are exactly as many geometries as there are subgroups of the projective group. Thus Euclidean geometry in a sense needs non-Euclidean geometry, because if this were missing, the projective group would be incomplete.

2.1.5. The Constitutive Form of Language of Synthetic Geometry

Klein's *Erlangen program* deepened our understanding of geometry. In contrast to Euclidean or Lobachevski's geometry, in which the concept of distance was a property of the space, for Klein the metrical structure ceases being something given a-priori, something unchangeable. It becomes a structure, which is introduced into the "neutral" projective space with the help of a transformation group. Nevertheless, this reconstruction still presupposes the a priori givenness of some geometrical object, namely of the projective plane. So the question arises whether we can overcome also this givenness, whether it is possible also to transform the projective plane, which in the *Erlanger program* represents the a priori presupposition of each metrical geometry, into an a posteriori construction. The answer to this question is positive. It is possible to make geometry free from the assumption of the a priori givenness of the projective plane. Nevertheless the way in which it can be done is much more radical than was Cayley's transition from the Euclidean to the projective plane. Here we are not going to some more fundamental kind of plane as a basis of geometry, in which the projective plane would be only one of the possible defined structures. The step made by Riemann consisted in giving up any kind of plane. His idea was to stop conceiving points, lines or planes as objects given in some space – projective or otherwise. Riemann's aim was to get rid of space completely.

2.1.5.1.   The Implicit Variant of the Constitutive Form – Riemann

Riemann realized the necessity to free geometrical objects from space in the course of his work in the theory of complex functions. A function of a complex variable is a function which prescribes to a point z of the complex plane Z a value $w = f(z)$ which is a point w of the complex plane W. The basic difficulty with such functions is that it is impossible to draw a graphical representation of them. The problem is that the domain and the range are both two-dimensional and thus the graph of a function of a complex variable, analogous to the sinusoid or logarithmic curve, would require a four-dimensional space. This is clearly beyond the limits of geometrical representation. However in the case when the function is one-to-one, i.e. to any two different values z_1 and z_2 there correspond two different values $w_1 = f(z_1)$ and $w_2 = f(z_2)$, it is possible to give a relatively clear picture about its behavior. We can draw two copies of the complex plane, one beside the other (of course, we are drawing only parts of them), and mark which regions of the plane Z are mapped onto which regions of the plane W. Unfortunately this approach is very limited. Even such simple functions as the power function $w = z^n$, the exponential function $w = e^z$ or the trigonometric functions $w = cos(z)$ and $w = sin(z)$ are not one-to-one. Riemann found an original way, in which they can be made one-to-one. It is enough to imagine that these functions are defined not on a complex plane, but on a more complicated object, which we get by pasting several copies of complex planes together.

Let us take for instance the function $w = z^2$. This function is not one-to-one because it maps one half of the complex plane Z onto the whole plane W. Thus the image of the whole plane Z covers the plane W twice. (For instance the points $z_1 = i$ and $z = -i$ are mapped onto the same point $w = -1$.) Nevertheless this double covering of the plane W is very regular. It is just as if two pictures were glued one on the other without any interfering or intermingling. Riemann's idea was to separate them. For this reason instead of one plane W on which the picture is depicted twice, we have to take two copies W_1 and W_2 of the plane W. On one of them we depict the upper half-plane of the plane Z (by which we shall cover this plane completely) and on the other we depict the lower half-plane of the plane Z. In this way we get rid of the overlapping of the two images. Our construction has nevertheless one fault. The image of the plane Z in the transformation $w = z^2$ should be "one piece". If the point z moves along a curve on the plane Z, its image w moves on the corresponding curve of the plane W. Contrary to this, our model consists of "two pieces". So every time that the point z passes from the upper half of the plane Z to its lower half, the point w "jumps" from the plane W_1 onto W_2. It is clear, that "in reality" there is no such jumping. It is merely an artifact of our model. We are confronted with jumps only because we divided what was originally unified.

Riemann found a way in which we can overcome this jumping. It is necessary to paste the two planes W_1 and W_2 together. (Of course not to paste them one onto the other – this is from where we started – but first to cut them and then paste them together along this cutting.) The boundary B which separates on the plane Z the upper half-plane from the lower one is the line, by the crossing of which the "jumping" from the plane W_1 onto W_2 occurs. The line B is mapped onto the positive halves of the real axes of the planes W_1 and W_2. So let us cut the planes W_1 and W_2 along these positive halves of their real axes and paste them together in such a way that no jumps will occur. We will

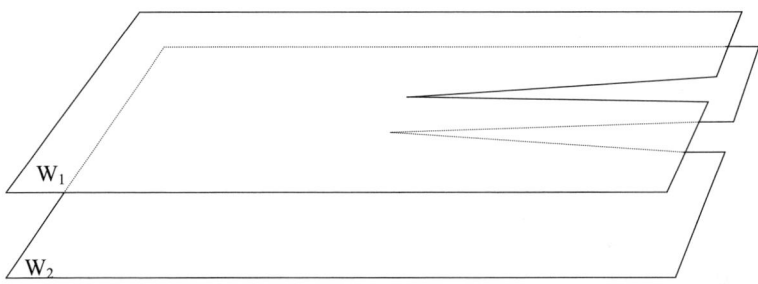

find out that we need to paste the upper edge of the cut of the plane W_1 to the lower edge of the cut of the plane W_2 and the lower edge of the cut of the plane W_1 to the upper edge of the cut of the plane W_2. After some trials we find that it is not possible. But why is it not possible? Because the space does not allow it. And here comes Riemann with a radical idea: *Let us forget about the space! Let us imagine that the pasting is done.* The function $w = z^2$ becomes in this way a one-to-one mapping of the complex plane Z onto the "double-plane" $W_1 W_2$ (which is an example of what we today call a Riemann surface). An analogous construction to that we did for the function $w = z^2$ can be done also for the other elementary functions of a complex variable mentioned earlier. In this way we acquire a geometrical insight into the behavior of these functions, which among others makes it possible to determine the values of many integrals without actual integration. We can guess the values of these integrals from the geometrical properties of the integral paths on the particular Riemann surface.

We presented Riemann's construction in its original context – the theory of functions of complex variables – because here emerged for the first time the idea *of considering a geometrical object indepen- dently of any space.* These original constructions were to some extent complicated by the circumstance that we had to paste together many copies of the complex plane (which is an infinitely extended object). Nevertheless we can show the basic steps of Riemann's construction on a much "simpler" object, on the so-called *Klein bottle.* Let us take a square and we are going to do what Riemann did with the complex planes, namely to paste together its edges. By every such pasting we have to determine, which two edges are we going to paste together and in which orientation. We will indicate with letters the corresponding edges and with arrows the orientation. If we paste two opposite sides,

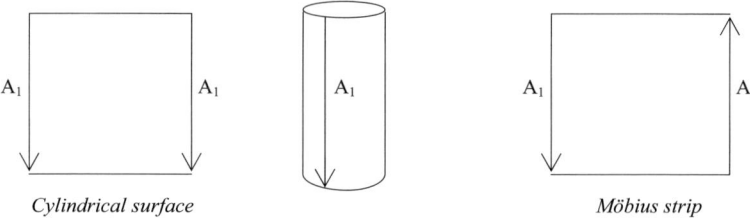

Cylindrical surface   *Möbius strip*

which are agreeingly oriented, we get the surface of a cylinder. If we paste two opposite sides of the square but one of them we twist 180 degrees, we get the well-known *Möbius strip.* Both these pastings are easy to perform in our three-dimensional space, though perhaps for the construction of the *Möbius strip* it is better to take a longer strip of pa- per or of cloth, rather than a square. But these are only technical details of secondary importance as the objects, at which we are aiming, will be impossible to construct, even with paper or cloth of any conceivable quality.

The next object, which is easy to construct, is the *torus*.

Let us now imagine a square, which differs from that defining the *torus* only in that we have changed the orientation of one of the edges A_2 to the opposite. We see that in this case it is impossible to paste the corresponding circles together, because their orientation does not fit. We would need to twist one of them, as we did it with one edge of the square in the construction of the *Möbius strip*. In the case of the *Möbius strip* we were lucky. The object before the pasting was

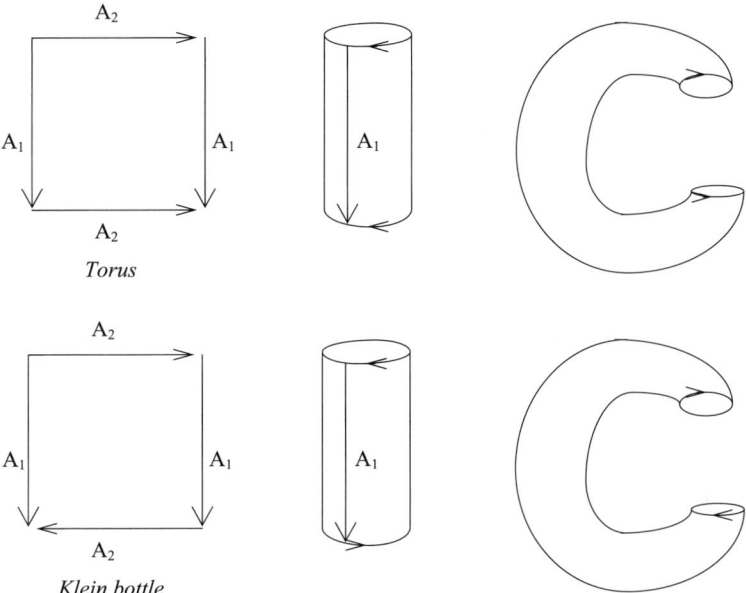

Torus

Klein bottle

planar and thus we had the third dimension of the space at our disposal, which we used for the twisting of one edge of the cut. In this way we obtained the correct orientation of the edges and so could paste them together. In the case of the *Klein bottle* we are in a worse position. The cylindrical surface, the one edge of which we have to twist, is a three-dimensional object, so we have no further free dimension at our disposal, into which we could "lean out" and make the necessary twist. But it is also clear that this is our problem, a problem of our three-dimensional space in which we want to do our construction. The *Klein bottle* itself has nothing to do with this. Whether we can or cannot construct it in the three-dimensional space is not a property of the object but of our space. Therefore we can consider as a new geometrical language the diagram consisting of a square (or more generally of a polygon with an even number of edges), on which it is indicated which edges, and in what orientation, should be pasted together. This language makes it possible to represent objects independently of their space. The basic advantage of the new language consists in the fact that it makes it possible to describe in a uniform way the surfaces which can be represented in our three-dimensional space and those which cannot. In this language we can show many interesting connections among surfaces, and this also in spite of the fact that we are not able to imagine

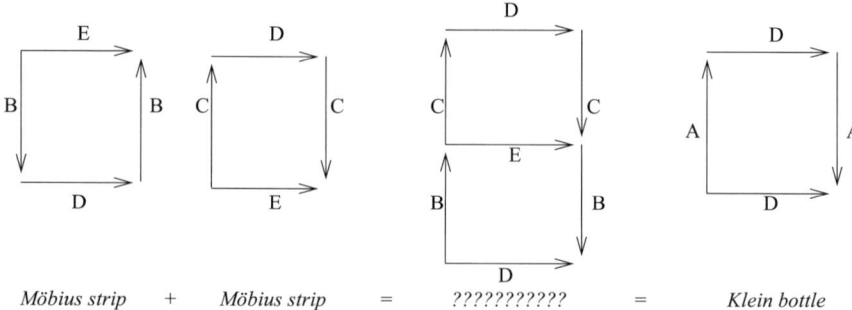

Möbius strip   +   Möbius strip   =   ??????????   =   Klein bottle

or represent the particular surfaces in our space. For instance, if we take a *Möbius strip*, we see that its edge is a circle. So let us take two copies of the *Möbius strip* and paste them together along these circles. Of course, we are unable to do this pasting. But neglecting this "detail", we can show that what we would get by this pasting, if we could perform it, is the *Klein bottle*. Thus we have found a relation between two objects, which we can not realize. Is this not nice? We see that

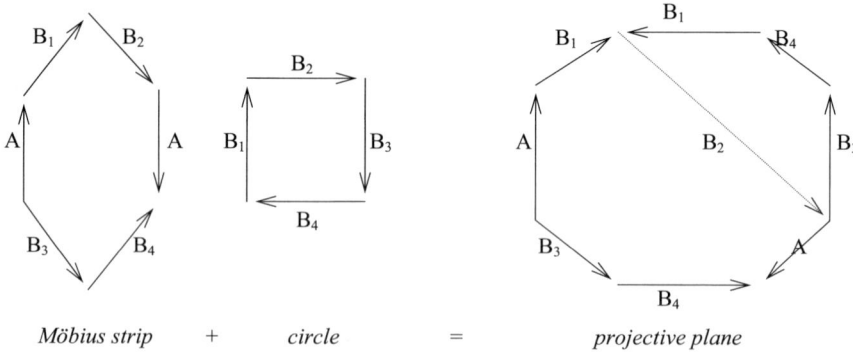

Möbius strip   +   circle   =   projective plane

Riemann's language, despite its apparent simplicity, is a powerful tool. Another example of a surface which it is not possible to construct in three-dimensional space, is the projective plane. We obtain it if we paste to a *Möbius strip* along its side (a circular line) a circle. Again, we are unable to carry out this pasting, because R^3 is too small and it is impossible to paste the particular circular lines (the edge of the *Möbius strip* and the edge of the circle) together.

> "Consider the problem faced by a 2-dimensional person living in R^2 who wants to form a closed surface from D^2 by pasting another disk D to D^2 along their common boundary.

He would be unable to do this in R^2 but a 3-dimensional person in R^3 could do it by making D into a 'cap'. The point of this story has been made before: *We should think of space more intrinsically and not as being imbedded in some special way in a particular Euclidean space..*" (Agoston 1976, p. 60)

But who can do this? I think that here we again have to do with an implicit appeal, which is similar to Lobachevski or Cayley. The only difference is that this time we are not required to see beneath a Euclidean triangle a non-Euclidean object, or beneath a Euclidean plane a projective one. Now we have to get rid of the space. From our previous experience with such appeals we know that also here we have to do with a new form of language, based on a new kind of epistemic subject. Our task is to clarify what kind of subject it is. So let us return to the picture representing the projective plane.

The reader found it perhaps strange that in the construction of the projective plane I was speaking about pasting of the *Möbius strip* to a *circle*, but I draw a *square*. But from the topological point of view the circle and the square are indiscernible objects, or more strictly speaking they are the same object. Topology considers two objects being equivalent (the technical term is homeomorphic) if we can get one from the other by a continuous deformation. It is easy to see that, by appropriate deformation of the sides of the square, we can get a circle. This aspect of topology shows that *topology is a language in which transformation groups are already incorporated.* Topology, exactly in the spirit of Klein, studies the invariants of a particular group of transformations. The only difference is that this transformation group is the group of homeomorphisms. The principle is nevertheless the same. The topological properties are defined as invariants of homeomorphisms. So it is clear that topology is based, from the very beginning, on Klein's integrative form of language.

Riemann's language contains something fundamentally new – cutting and pasting. These operations lie beyond the boundaries of the *Erlanger program.* If we were by chance to cut the projective plane, we would destroy the projective group and so the whole structure of Klein's language as well. This means that cutting and pasting are more radical operations than the projective transformations. If we consider a square or a circle (from the topological point of view they are the same), then to the inside of this circle there corresponds a group of transformations. If we paste the two opposite sides of the square together and construct in this way a cylindrical surface, we change the transformation group. The new group will contain, besides the transformations of the circle also the transformations corresponding to the rotations of the cylindrical surface round its axes. If we create from the cylindrical surface a *torus*, we again enrich the transformation group. So if we consider small creatures living on the cylindrical surface or on a *torus*, if they would formulate some day something analogous to the *Erlanger program* for their world, as a basis for it they should take not the group of the projective plane, as Klein did, but the group of transformations of the cylinder or of the *torus*.

What I want to say is that each of these surfaces has its own geometry, corresponding to a particular transformation group. The Riemannian cutting and pasting crosses these geometries and makes it possible to pass from one of them to another. Riemann's language describes the constitutive acts with the help of which we can create the surfaces, on which the whole Kleinian apparatus of transformation groups works. Therefore the form of language, which forms the basis of Riemann's language, is the *implicit constitutive form*. It is implicit because Riemann is not able to tell us what exactly we have to do in order to get rid of the three-dimensional space and to see the *Klein bottle* or the projective plane. And it is constitutive because it takes over the role, which geometry until Riemann had assigned to space – the role of constituting the existence of objects. Riemann has found how we can grasp this constitutive function of space. In this respect he resembles Cayley and his grasping of the metrical structure of the Euclidean plane. In both cases something which was given, which represented the a-priori characteristics of the space, is replaced by an a-posteriori structure.

2.1.5.2.   *The Explicit Variant of the Constitutive Form – Poincaré*

The basic problem of Riemann's language was the implicit nature of its form. On the one side it relies on geometrical intuition (we have to see that by pasting together the sides A_1 and A_2 we get a *torus*), but on the other side it requires us to abandon a great part of this intuition (namely its dependence on three-dimensional space). The way out from this dilemma was found by Poincaré. It is the well-known combinatorial topology that brought an incorporation of the implicit constitutive subject into the language. In a similar way to our proceeding in the case of Beltrami's model, here also we will not follow in detail the historical development. I will not present Poincaré's original version of combinatorial topology, but only its simplified version, developed by Brouwer.

But for the sake of simplicity I will not distinguish between them and will speak of Poincaré, where I should really speak about Brouwer.

If we want to represent the projective plane, we will not mention any pasting (which we are unable to carry out), and also we will not require from the reader to forget about space (which he is unable to do). Let us take a picture of Riemann's language and divide it into triangles. It may seem that we were too generous in this triangulation and have chosen too many triangles. But the aim is to make sure that no two segments which correspond to different edges have the same labels. It is obvious that the internal hexagon $(v_4 v_5 v_6 v_7 v_8 v_9)$ represents a circle, while the external belt represents a *Möbius strip*, and thus the whole object is a projective plane.

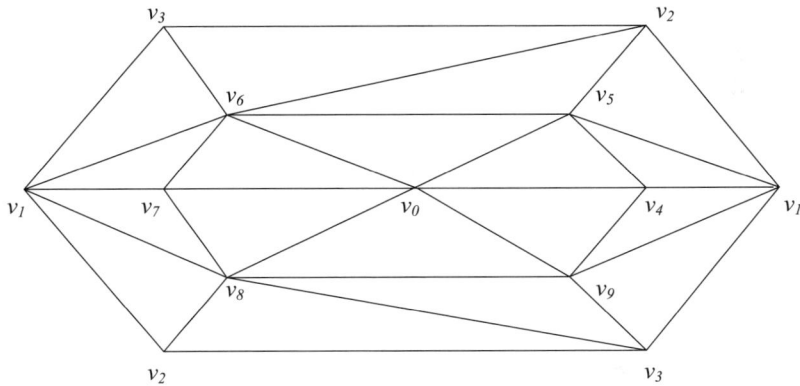

Now we define a k-dimensional simplicial complex (see Agoston 1976, p. 32):

Definition 1:   *Let $k \geqslant 0$. A k-dimensional simplex is the convex hull σ of $k + 1$ linearly independent points $v_0, v_1, \ldots, v_k \in R^n$. We write $\sigma = v_0 v_1 \ldots v_k$. The points v_i are called the vertices of σ.*

Definition 2:   *Let $\sigma = v_0 v_1 \ldots v_k$ be a k-dimensional simplex and let $\{w_0, w_1, \ldots, w_l\}$ be a nonempty subset of $\{v_0, v_1, \ldots, v_k\}$, where $w_i \neq w_j$ if $i \neq j$. Then $\tau = w_0 w_1 \ldots w_l$ is called an l-dimensional face of σ.*

The point v_0 is a 0-dimensional simplex, the segment $v_0 v_1$ is a 1-dimensional simplex and the triangle $v_0 v_1 v_2$ is a 2-dimensional simplex. Thus the simplexes represent the simplest objects of the particular dimension. The simplexes v_0 and v_1 are 0-dimensional faces and

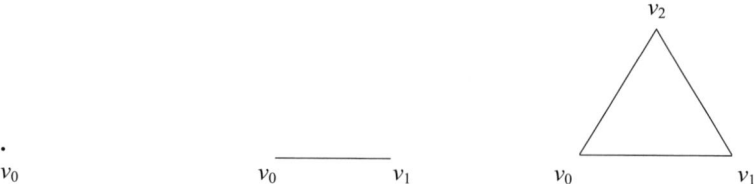

the v_0v_1 is a 1-dimensional face of the simplex v_0v_1. The simplex v_0v_1 has no other faces. The simplex $v_0v_1v_2$ has three 0-dimensional, three 1-dimensional and one 2-dimensional faces.

Definition 3:   *A simplicial complex K, is a finite collection of simplices in some R^n satisfying:*

    1. *If $\sigma \in K$, then all faces of σ belong to K*

    2. *If $\sigma, \tau \in K$, then either $\sigma \cap \tau = \emptyset$ or $\sigma \cap \tau$ is a common face of σ and τ.*

For instance the sphere (i.e. the surface of a ball), which from the topological point of view is an object equivalent to the surface of the tetrahedron, we can represent as the following simplicial complex:

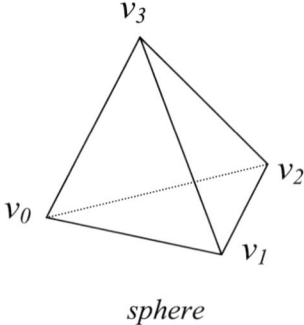

sphere

$$K_S = \{v_0v_1v_2, v_0v_1v_3, v_0v_2v_3, v_1v_2v_3, v_0v_1, v_0v_2, v_0v_3, v_1v_2, v_1v_3, \\ v_2v_3, v_0, v_1, v_2, v_3\}.$$

Similarly for the projective plane we get the complex

$$K_P = \{ v_1 v_6 v_3, v_2 v_3 v_6, v_2 v_6 v_5, v_1 v_2 v_5, v_1 v_5 v_4, v_7 v_0 v_6,$$

$$v_1 v_9 v_3, v_3 v_9 v_8, v_2 v_3 v_8, v_1 v_2 v_8, v_1 v_8 v_7, v_1 v_7 v_6, v_1 v_4 v_3,$$

$$v_6 v_0 v_5, v_5 v_0 v_4, v_0 v_9 v_4, v_8 v_9 v_0, v_7 v_8 v_0, v_1 v_2, v_1 v_3, v_1 v_4,$$

$$v_1 v_5, v_1 v_6, v_1 v_7, v_1 v_8, v_1 v_9, v_2 v_3, v_2 v_5, v_2 v_6, v_2 v_8, v_3 v_6,$$

$$v_3 v_8, v_3 v_9, v_4 v_5, v_4 v_9, v_4 v_0, v_5 v_6, v_5 v_0, v_6 v_7, v_6 v_0, v_7 v_8,$$

$$v_7 v_0, v_8 v_9, v_8 v_0, v_9 v_0, v_1, v_2, v_3, v_4, v_5, v_6, v_7, v_8, v_9, v_0 \} \ .$$

Now we can forget about Riemann's picture of the projective plane. The simplicial complex K_P represents the projective plane without any reference to pasting. The edge $v_1 v_2$ in the triangles $v_1 v_2 v_5$ and $v_1 v_2 v_8$ is simply *the same edge*. In the picture it is present twice, once at the right lower corner and the second time at the left upper corner. But this is the problem of the picture. It is impossible to draw the projective plane without cutting it and thus the edges which form the cut will be in the picture twice. On the other hand the simplicial complex represents the particular surface without any reference to cutting or pasting. Thus in the complex every vertex, edge and face of the projective plane is present only once (as it is listed only once in the definition of K_P).

In this way we have got rid of the picture. But we have not got rid of the space. This is because until now we have defined only the so-called concrete simplicial complex, that is a complex which "dwells" in some space R^n (see Definition 1). Our next step is thus to separate the language of the simplicial complexes from reference to any space. We reach it so that we deprive the above listed system of symbols (such as K_P) of their geometrical interpretation. For this reason we define the so-called abstract simplicial complex, which will be just a system of symbols. The symbols which we presented in the case of the sphere (K_S) or the projective plane (K_P) contain everything that is important about the sphere or the projective plane. All topological properties such as simple connectedness, dimension or Euler characteristic can be calculated using only the symbols, i.e. without any reference to a picture or space.

The basic achievement of Poincaré was the development of formal techniques, which only on the basis of the abstract simplicial complexes make it possible to determine the basic topological invariants of a surface. Without any reference to a picture or space, using only symbols, he managed to compute the Betti numbers (which determine connectedness, dimension, Euler characteristic and many other topological invariants). *Every surface is constituted by the way in which the triangles of its triangulation are connected.* We need no other information in order to determine the topological invariants, and exactly this information is expressed in the abstract simplicial complex. Thus Poincaré succeeded in incorporating Riemann's constitutive acts into the language. Instead of pasting the edges of a square in our imagination (as the real pasting is impossible to carry out, it remains our task to pretend that we know what would come out in such pasting) we make the particular triangulation, in which the pasting is already fulfilled (or, more precisely, there is nothing to paste, because nothing was cut). The language of combinatorial topology[9] and topology in general is thus based on the *explicit constitutive form of language*. It is the language by which mathematics liberated itself from dependence on space. It makes it possible to speak about objects independently of whether it is or it is not possible to represent them in our three-dimensional space. The simplicial complex of the projective plane does not differ in any fundamental way from the complex, which corresponds to the sphere. The fact, that it is not possible to realize it in our space is only of secondary importance.

2.1.6. The Conceptual Form of Language of Synthetic Geometry

One of the most important discoveries of Georg Cantor was the realization of the fact that it is possible to present all constitutive acts in a uniform way as the grasping of a particular system of objects as one whole. Cantor called such systems *sets*. For instance a simplicial complex is nothing else than grasping the particular simplexes into one complex, and so it is a set of simplexes. The form of language that emerged after the constitutive form was based on the set-theoretical approach. If we

---

[9] The second branch of topology – called general topology – has a similar form of language, as combinatorial topology. A topological space – the central notion of general topology – is a set on which a topology is introduced. Topology can be introduced in different ways: using open sets, closed sets, or a filter. But what is common for all these ways is that *the structure of closeness* (i.e., the determination of which points are close to each other) *is constituted by the language*. Before the creation of general topology, closeness was determined by space. Two points were considered as close or remote depending on their position in space. In general topology *closeness is liberated from space*. Similarly to combinatorial topology also here space stops being the constitutive basis of language. Closeness ceases to be a relation constituted by space, which only "lends" it to its points. The constitutive role that was formerly played by space is taken by language. Therefore the notion of a topological space is from an epistemological point of view closely related to the notion of a simplex from combinatorial topology.

realize that this is the basic function of concepts – to make it possible to grasp systems of objects – we can call the form of language that is based on the set-theoretical approach the *conceptual form of language*.

2.1.6.1.   The Implicit Variant of the Conceptual Form – Hilbert

One of the creators of set theory, Richard Dedekind, introduced in his paper *Continuity and irrational numbers* (Dedekind 1872) the construction of real numbers as *cuts*. These cuts are, according to Dedekind, sets of rational numbers that fulfill specific conditions. Dedekind did not use the term set. Instead he spoke about *systems*; nevertheless, he used the term system in the same sense as Cantor used the term set. Dedekind's construction of the real numbers has all the features characteristic of the *constitutive* form of language: the new objects (real numbers) are created by means of *constitutive acts* (Dedekind even directly says that by means of a cut we create a new irrational number). These constitutive acts consist in grasping a particular system of objects (a system of rational numbers) *as one whole*. Thus Dedekind's cuts are from the epistemological point of view analogous to Poincaré's simplicial complexes. Nevertheless, this aspect of Dedekind's paper is now secondary. From the point of view of the development of geometry, Dedekind's paper contains an important passage discussing the continuity of the real line:

> "I am glad if every one finds the above principle so obvious and so in harmony with his own ideas of line; for I am utterly unable to adduce any proof of its correctness, nor has any one the power. The assumption of this property of the line is nothing else than an axiom by which we attribute to the line its continuity, by which we find continuity in the line. If space has at all a real existence it is not necessary for it to be continuous; many of its properties would remain the same even were it discontinuous.." (Dedekind 1872, pp. 11–12)

This passage indicates that the continuity of a straight line or a circle, which Euclid assumed, is not automatically guaranteed. A plane formed of points having both co-ordinates rational numbers is a model for Euclid's axioms. Nevertheless, in such a model there can be two circles having the sum of their radii greater than the distance of their centers and the difference of their radii smaller than this distance, and despite this they would have no intersection. The reason is that the candidate for the point of their intersection could have one or both of its co-ordinates irrational, and so it would not belong to the model. Euclid assumes in several of his proofs the existence of points of intersection of such circles. The existence of these points (as our model of the Euclidean plane indicates) is not guaranteed by the axioms of Euclid. Thus it seems that Euclid simply assumed it on the basis of intuition.

Shortly after the publication of Dedekind's paper, two important works on the foundations of geometry appeared: *Vorlesungen über neuere Geometrie* (Pasch 1882) and *Grundlagen der Geometrie* (Hilbert 1899). They both filled in the above mentioned gap in Euclid by adding new axioms to the Euclidean system. In the case of Hilbert, among the new axioms were the Archimedean axiom and the axiom of linear completeness:

> **Archimedean axiom.** Let A_1 be any point upon a straight line between the arbitrarily chosen points A and B. Take the points $A_2, A_3, A_4 \ldots$ so that A_1 lies between A and A_2, A_2 between A_1 and A_3, A_3 between A_2 and A_4, etc. Moreover, let the segments
>
> $$A A_1, \quad A_1 A_2, \quad A_2 A_3, \quad A_3 A_4, \ldots$$
>
> be equal to one another. Then, among this series of points, there always exists a certain point A_n such that B lies between A and A_n.
>
> **Axiom of completeness.** To a system of points, straight lines, and planes, it is impossible to add other elements in such a manner that the system thus generalized shall form a new geometry obeying all the five groups of axioms. In other words, the elements of geometry form a system which is not susceptible of extension, if we regard the five groups of axioms as valid. (Hilbert 1899, p. 15)

After adding these axioms, Hilbert proved the consistency and the mutual independence of the axioms of his system. All his proofs were strictly logical; the whole argumentation was based solely on principles that were explicitly stated in the axioms. In this way geometry was for the first time *separated from spatial intuition and from any a priori interpretation of its fundamental categories.*

In the *constitutive* form of language, the language took over the function of constituting the existence of geometrical objects and thus made it possible to study objects for which ordinary three-dimensional space was too small. Such an object is for instance the projective plane.

But despite the explicit constitution of the existence of the objects, the mathematical work with these objects and the proofs of theorems about them was still to a considerable degree intuitive. The language constituted only the existence of objects; the essence of them was still more or less intuitively given. Thus for instance the projective plane remained what it always was; its essence and its properties did not change. Only its existence was put on more solid foundations by use of the particular simplicial complex. In the constitutive form of language the objects preserved their natural properties.

With Hilbert a new form of language entered the scene, the form in which an object has those and *only those* properties that are introduced by the axioms. Thus not only the existence but also *the essence of the objects starts to be constituted by the language.* In this way geometrical objects become separated from their natural properties that were given in intuition. For Hilbert it is unimportant what we have in mind when we formulate our theories. In the proofs we are allowed to use only those properties which we introduced in the axioms and definitions. So with Hilbert begins the expelling of intuition from the foundations of geometry. I suggest calling the new form of language used by Hilbert the *conceptual form of language.* It is the form on the basis of which the conceptual structure of mathematical theories is separated (pruned or purified) from intuition. Concepts played a fundamental role also at the previous stages of the development of geometry, but those concepts were hopelessly intertwined with intuition. (How hopeless this intertwining was is illustrated by the gaps in Euclid concerning the notion of continuity, which were discovered by Dedekind. For more than two millennia nobody realized these conceptual gaps, because Euclid's proofs were so intuitively persuasive.)

2.1.6.2. *The Explicit Variant of the Conceptual Form – Tarski*

Hilbert's system of axioms is based on the implicit variant of the conceptual form of language. Implicit because Hilbert remained faithful to some fundamental principles of the Euclidean tradition and in a sense only filled in the gaps of the Euclidean system. Thus for instance Hilbert preserved Euclid's assumption about the existence of different kinds of fundamental objects. Even though he separated these objects from their traditional connection to intuition, he preserved their difference (which has its origin in our intuition) when he started his *Grundlagen der Geometrie* with the passage:

"Let us consider three distinct systems of things. The things composing the first system, we will call points and designate them by the letters $A, B, C, \ldots$; those of the second, we will call straight lines and designate them by the letters $a, b, c, \ldots$; and those of the third system, we will call planes and designate them by Greek letters $\alpha, \beta, \gamma, \ldots$ We think of these points, straight lines, and planes as having certain mutual relations, which we indicate by means of such words as "are situated", "between", "parallel", "congruent", "continuous", etc. The complete and exact description of these relations follows as a consequence of the axioms of geometry. These axioms may be arranged in five groups. Each of these groups expresses, by itself, certain related fundamental facts of our intuition." (Hilbert 1899, p. 2)

In the phrase *"let us imagine three different systems of things"* at the beginning of the passage, the word system designates a set and thus Hilbert's system uses the set-theoretical approach. But it uses it only to a certain degree, because Hilbert understood straight lines and planes in a way similar to Euclid, as independent objects and not as sets of points. That means that Hilbert preserved certain aspects of the traditional understanding of geometrical objects. For transition to the explicit version of the conceptual form of language, transformation of the straight lines into sets of points was important. At the same time space, which was traditionally understood as emptiness into which the particular geometrical objects were placed, was now turned into a set of points. In this way space became something full; it was filled with points. The transformation of the continuum into a point set occurred in the second half of the nineteenth century under the influence of the theory of functions of a real variable.[10] It is likely that this idea was introduced into geometry only in the work of Tarski.

As a turning point that indicates the transition to the *explicit variant of the conceptual form of language* of synthetic geometry we can take Tarski's work *The completeness of elementary algebra and geometry*. As Alonzo Church mentioned in his review (Church 1969), the results contained in this work were obtained by Tarski around 1930. The manuscript of the book was accepted for publication in 1939 by the publisher Hermann & Cie. in Paris. Work on the manuscript reached

---

[10] The transformation of the continuum into a set is described in (Medvedev 1965).

the stage of proofreading, but as a result of war activities the whole typesetting was destroyed and only two exemplars of the printed book survived. After the war Tarski rewrote the whole book and published it under the title *A decision method for elementary algebra and geometry* (Tarski 1948).

In contrast to Hilbert who has three kinds of objects (points, straight lines, and planes) Tarski had only a single kind of objects – points (thus he considered straight lines as particular sets of points). Similarly, instead of the five Hilbertian relations among objects, Tarski introduced only two relations – a ternary relation of lying between and a quaternary relation of being equidistant. This makes it possible to formalize Tarski's system in the framework of a (one sorted) first-order predicate calculus and to study it using the methods of model theory. Tarski succeeded in proving deep results – the completeness and decidability of his axiomatic system of elementary geometry. This result seems to be an appropriate point to end our exposition of the development of synthetic geometry.

2.1.7. An Overview of Relativizations in the Development of Synthetic Geometry

We distinguished six different forms of language in the development of synthetic geometry, i.e. six ways in which it is possible to connect the language of geometry with its universe. In the first of them, called *perspectivist form of language,* the language represented objects from a particular point of view and so it was able not only to represent the isolated objects themselves but also their relative positions in space. Everything else remained synthetic, i.e. given by extra-linguistic means, in the case of geometry that means by intuition. Gradually more and more aspects of geometrical intuition were incorporated into the language, and in the *conceptual form* the language constitutes all attributes of objects. In this gradual evolution geometry is more and more separated from intuition and we could say that the intuitive aspects of geometrical objects are gradually incorporated (reified, objectified) in the structure of language. In the end the discipline which was for centuries the paradigm of synthetic principles was turned into a purely formal axiomatic system which turned out to be decidable. This span which geometry passed on its way from Euclid to Tarski is expressed in the following scheme, which summarizes the results of our analyses. In the scheme the different forms of language are associated to two opposite poles, between which the evolution of the form of language weaves in the form of a zigzag. This aspect of the scheme will be explained later.

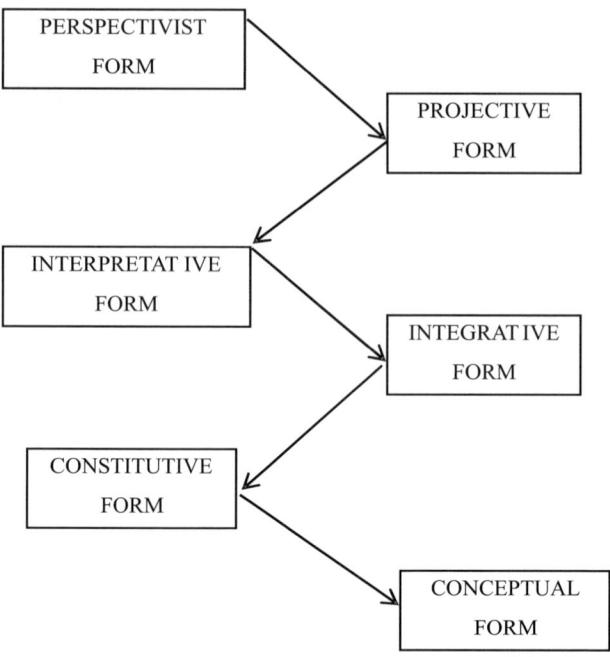

2.2. Historical Description of Relativizations in Algebra

While geometry represents a classical topic in the philosophy of mathematics – it suffices to mention Plato, Pascal, Spinoza, Kant, or Husserl – the philosophical questions posed by algebra have remained on the periphery of interest. Apart from the books of Jacob Klein (Klein 1934) and Jules Vuillemin (Vuillemin 1962), little attention has been paid to algebra in the philosophical literature. This neglect of algebra in the philosophy of mathematics has many reasons. One of them could be the dominant influence of Ancient Greek culture on the formation of Western philosophical tradition. The Greeks approached the analysis of human thought mainly on the basis of the metaphor of visual experience.[11] Algebra, which has Arabic roots and is closer to motoric than visual experience, could not find in Western philosophical tradition an adequate conceptual framework for the analysis of changes occurring in its history. Therefore the changes in the development of algebra remained outside of the realm highlighted by philosophical reflection. The aim of the present chapter is to describe the fundamental changes in the development of algebraic thought and to reflect on their philosophical significance. To this end I will use the approach that I introduced in the analysis of the development of geometry. In particular I will divide the development of algebra into distinct periods using terms such as *"perspectivist form of language"* or *"projective form of language"*.

The reader may be surprised that just after I have stressed the differences between algebra and geometry I am trying to transfer the periodization from the history of geometry to the history of algebra. Nevertheless, this contradiction is only an apparent one, because already the periodization of the history of geometry was based on the analysis of its language. Geometry was used as the material by the analysis of which I studied the changes in the epistemological structure of the language of mathematics. The choice of geometry as a point of departure for the analyses of the development of mathematics was natural. In Western philosophical tradition, geometry is perhaps the most deeply reflected and understood mathematical discipline. But after I found some regularity in the development of geometry, it seemed natural to try to use it also in the interpretation of the development of algebra. This explains the periodization of the history of algebra based on notions such as *"perspectivist form of language"* or *"projective form of language"*.

---

[11] The following analyses are based on a confrontation of the role of visual and motor experience in the development of culture. When we say that in Ancient Greek civilization visual experience was dominant, we do not mean vision in the *physiological* sense, because that is more or less the same for all civilizations. It is a biological property common to all humans. In this respect there is no difference between an Ancient Greek and a medieval Arab. Nevertheless, it is a peculiarity of man that besides his physical surroundings he is surrounded also by cultural artifacts like myths, heroes, gods, values, etc. To these cultural artifacts belong also mathematical objects such as numbers, triangles, or algebraic equations. These objects transcend our physical surroundings – it is impossible to touch them. And here we meet again vision, but this time in a *cultural* sense, as one of the metaphors of approaching the transcendent reality. I believe that one of the fundamental differences between Greek and Arabic cultures consists in the circumstance that Ancient Greeks understood the transcendent reality as a kind of a parallel world which we can reach by vision (it suffices to recall Plato's metaphor of the cave). In contrast to this, for an Arab the road to transcendent reality was not based on the metaphor of sight but on the metaphor of hearing (of a divine voice that gives us orders). The transcendent reality was therefore not some parallel world but it was rather an alternative way of acting.

The notions of perspective or projection are doubtlessly geometrical notions. Nevertheless, I use them in an epistemological sense to characterize a particular form of language. For the perspectivist form, for instance, it is typical to consider expressions of language as pictures of reality seen from a particular point of view, while the projective form introduces the representation of a representation as a fundamental innovation. In geometry the transition from the perspectivist to the projective form is obvious. I would like to show that from the epistemological point of view a very similar transition also occurred in the development of algebra. At the beginning mathematicians accepted only a unique positive value of the unknown that fulfilled all the conditions stated in formulation of the mathematical problem as a solution of an equation. The unknown was thus seen as a "picture" of a quantity of the real world, precisely as in the perspectivist representation. Later, when mathematicians introduced substitutions[12] (i.e., representations of representations) the view arose that all roots of an equation are to be accepted as solutions.[13] This meant that not only the "*true*", that is, positive solutions, those with a direct relation to reality, were accepted but also (to use Descartes' term) the "*false*" and even the "*imaginary*" ones. The solution thus ceased to be a "picture" of the extralinguistic reality and became an element of a structure. Substitutions in many respects resemble the projections of projective geometry. Both introduce a representation of a representation. Just as the introduction of projections into geometry led Desargues to complement the plane by infinitely remote points in order to make projections one-to-one mappings, so in algebra it turned out to be necessary to complement the number system by negative and complex numbers in order to ensure appropriate closure properties. Thus we see an analogy between projections of a geometric figure and transformations of an algebraic equation.

---

[12] A substitution in algebra is an operation in which for one variable, for instance x, we substitute another expression, for instance $(y-1)$. An equation that using the variable x had the form $x^2+2x-8=0$, will after this substitution become $y^2-9=0$. It is easy to see that its solution is $y=3$. From this we can find the solution of the original equation $x=2$, because $x=y-1$.

[13] A root of an algebraic equation is any number that fulfills the equation. Thus for instance the equation $x^2+2x-8=0$ has two roots: $x=2$ and $x=-4$. Already our simple example indicates that an equation can be satisfied by several numbers. In the early stages of the development of algebra there was a tendency to reserve the term *solution* for that number, which satisfied the real situation that corresponded to the equation. In most cases this meant that it had to be positive. The other numbers that fulfilled the equation were called false solutions, or simply roots

When we transfer the notion of the form of language from geometry to algebra we discover that objects like 0 or 1 play a role in algebra similar to that played in geometry by notions like center of projection or horizon. The purpose of the form of language is to integrate the epistemic subject of the language into the universe of the language. In geometry, where the universe of the language lies before us and the epistemic subject has the form of a point of view, the position of this subject in the universe is fixed by means of the horizon. The horizon indicates the horizontal plane of our sight. In algebra, when compared with geometry, the situation is rather the opposite. The universe of algebra is created through reification of the activities of the epistemic subject. The numbers 3 or $\sqrt{5}$ are symbolically reified acts of counting or of extracting the root. Thus the universe of the language of algebra is not a region opened to our gaze where our only task is to fix our position. On the contrary, the universe of algebra is something that was originally part of ourselves, namely, of our actions, and we only distance ourselves from its objects step by step as we represent actions like counting, adding, subtracting, etc. by symbols of numbers, addition, or subtraction.[14] In this way the reification of algebraic operations like addition or root extraction brought the deeper structure of symmetries to the fore, which was in its turn reified in the form of a group.

The universe of algebra is thus not an external world open to our gaze. The universe of algebra arose in the process of gradual reification of linguistic descriptions of our activities. It is no accident that algebra was created by the Arabs rather than the Greeks. In our reconstruction of the development of algebra we will consider the history of algebra in western civilization. We will follow the history of western civilization, a civilization of the "geometric spirit", a civilization for which to understand means to acquire an insight, and we will reconstruct how this civilization integrated algebra, a discipline, the universe of which is not open to our gaze, a discipline that is based on the motoric experience, on manipulations, on actions. In contrast to the ideal of the Greek the-

---

[14] In this metaphor I consider language not as a rigid structure but as a moving organism. The metaphor is based on the fact that language has also a performative aspect that can be likened to the movements of the body. Writing brings a reification of this performative aspect – the performance is exempted from the flux of time. A written text resembles traces in sand and the development of the algebraic symbolism can be likened to the sloughing of the skin of a snake, where beneath one layer of skin grows the next one. The traces in sand are left there by the particular layer of skin. The movements of the body, the performance, the flux of transformations are primary. The symbolism is only dead skin, something that is created from the living body of algebraic thought in different layers of its reification.

oretician, who does not intervene in the events of the world and only follows them with his gaze, an algebraist always acts, performs calculations, solves equations, transforms expressions. In algebra, insights are possible only after an action, only after some lucky trick has brought us to the sought solution. But besides such positive experiences, when we find the solution, in algebra very often we have to deal with a negative experience, with the experience that the transformation did not lead to the desired result. Such negative experiences also require conceptualization; they lead to the desire to understand why the actions did not succeed. One of the central threads in the development in algebra was motivated by the wish to understand the systematic failure of all attempts to solve the quintic equation. Why is it that despite the attempts of the best mathematicians during the period of three centuries nobody managed to solve such simple equations as $x^5 - 6x + 3 = 0$?

In situations of this kind, the problem is not to arrive at an in *sight*, because at the beginning there is nothing on which our *sight* could rest. There is only a long line of failures, heaps of paper filling the wastebaskets of leading mathematicians – all this in thousands of variations. The drama of the development of algebra lies in the slow process by which mathematicians finally succeeded in determining the outlines of the invisible barrier blocking all attempts to solve the quintic equation. This barrier is called the alternating group of five elements. The mathematicians who succeeded in determining its outlines were Rufffini, Abel, and Galois.

Algebra with its stress on manipulation of formulas was alien to the Greek spirit. Algebraic manipulations are not based on theoretical insight, but on skill in calculations. The aim is not to envision the result, but rather to acquire a sense for the tricks that can lead to it. Algebra is always about a process. When Arabic algebra came through Spain to Europe, a dialogue started between the spirit of western mathematics and this fundamentally different, but equally deep, spirit of algebra. This dialogue was dominated by attempts to visualize, the endeavor to get all the tricks and manipulations into the visual field and to acquire an insight into them. But whenever the western spirit succeeded in visualizing one layer of algebraic thought, another layer emerged beneath it. In our reconstruction we will try to show how, beneath the rules of algebra, algebraic formulas emerged, beneath the algebraic formulas forms, beneath the forms fields, and finally beneath the fields groups. The history of algebra is a history of a gradual reification of activities, a history of transforming operations into objects.

2.2.1. The Perspectivist Form of the Language of Algebra

Abú Abdalláh Muhammad al-Khwárizmí is the author of the *Short book of al-jabr and al-muqábala*, a treatise on solving "equations". The word *al-jabr* in the title of the book became in time to be used as a name for the whole discipline dealing with "equations". We cannot speak about equations in the modern sense, because the book of al-Khwárizmí, remarkably, makes no use of symbols, and even numbers are expressed verbally. For the powers of the unknown the book uses special terms: for x it uses *shai* (thing), for x^2 *mál* (property), for x^3 *kab* (cube), for x^4 *mál mál*, for x^5 *kab mál*, etc. In translations of the work the Arabic names for the powers of the unknown were replaced by their Latin equivalents; thus *res* stood for *shai*, *census* for *mál*, and *cubus* for *kab*. In Italy the word *cosa* replaced the Latin *res*, and so during the fifteenth and sixteenth centuries algebra was usually named *regula della cosa* – the rule of the thing. Algebra was understood as a set of rules for manipulating the thing (i.e., the unknown), which enable us to find the solutions of particular "equations".

Before attempting to solve a particular "equation", al-Khwárizmí first rewrote it in a form where only positive coefficients appeared and the coefficient of the leading term (term with the highest power of the unknown) was 1.[15] In order to achieve this form, he made use of three operations: *al- jabr* – if on one side of the "equation" there are members that have to be taken away, the corresponding amount is added to both sides; *al-muqábala* – if the same power appears on both sides, the smaller member on the one side is subtracted from the greater one on the other side; and *al-radd* – if the coefficient of the highest power is different from 1, the whole "equation" is divided by it. We write the term "equation" in quotation marks because, strictly speaking, al-Khwárizmí did not write any equations. Rather, he transformed relations among quantities, everything being stated in sentences of ordinary language, enriched by a few technical terms.

We can illustrate his approach with an example. Consider the equation $x^2 + 10x = 39$, which he expressed in the form: "*Property and ten things equals thirty-nine*". His solution reads as follows: "*Take the half of the number of the things, that is five, and multiply it by itself, you obtain twenty-five. Add this to thirty-nine, you get sixty-four. Take*

---

[15] For instance, he transformed the equation $2x^2 - 4x + 8 = 0$ into the form $x^2 + 4 = 2x$.

the square root, or eight, and subtract from it one half of the number of things, which is five. The result, three, is the thing." This approach is close to the Babylonian tradition. It is a set of instructions telling us how to find the solution. Nevertheless, there is a substantial difference. In contrast to the Babylonian mathematicians al-Khwárizmí has *the notion of the unknown (shai)* and therefore his instruction *"take the half of the number of the things, multiply it by itself, add this to thirty nine, take the square root, and subtract from it one half of the number of things"* is a universal procedure, which can be applied to any quadratic equation of that particular form. Thus, while the Babylonian mathematicians only made their calculations with concrete numbers, al-Khwárizmí is able to grasp the procedure of solution in its entire universality. When he uses concrete numerical values for the coefficients, he does so only for the purpose of illustration. With the help of the notions as *shai, mál,* and *kab* he is able to grasp the universal procedure, and in taking this step he became the founder of algebra.

In the twelfth century the works of al-Khwárizmí were translated into Latin. The custom of formulating the solution of an "equation" in the form of a verbal rule persisted until the sixteenth century. The first result of western mathematics that surpassed the achievements of the Ancients was formulated in this way. This was the solution of the cubic equation, and was published in 1545 in the famous *Ars Magna Sive de Regulis Algebraicis*[16] by Girolamo Cardano. Cardano formulated the equation of the third degree in the form: *"De cubo & rebus aequalibus numero."* The solution is given in the form of a rule: *"Cube one-third of the number of things; add to it the square of one-half of the number; and take the square root of the whole. You will duplicate this, and to one of the two you add one-half the number you have already squared and from the other you subtract one-half the same. You will then have* binomium *and its* apotome.[17] *Then subtracting the cube root of the apotome from the cube root of the binomium, that which is left is the*

---

[16] The works by which early modern science left behind the ancient tradition appeared almost simultaneously. In 1543 Copernicus published his *De Revolutionibus Orbium Coelestium* that introduced a revolution in our view of the universe, and Vesalius published his *De Fabrica Humani Corporis* that brought about a radical change in anatomy. Two years later, in 1545, Cardano published his *Ars Magna Sive de Regulis Algebraicis* that was a breakthrough in algebra.

[17] Cardano called $\frac{c}{2} + \sqrt{\left(\frac{c}{2}\right)^2 + \left(\frac{b}{3}\right)^3}$ a *binomiom* and $-\frac{c}{2} + \sqrt{\left(\frac{c}{2}\right)^2 + \left(\frac{b}{3}\right)^3}$ an *apotome* These terms can be found already in Euclid in the theory of proportions. Cardano never wrote formulas. I took the liberty of transcribing his verbal rules into modern algebraic notation.

thing." In order to *see* what Cardano was *doing*, we present the equation in modern form $x^3 + bx = c$ and we express its solution using modern symbolism:

$$x = \sqrt[3]{\frac{c}{2} + \sqrt{\left(\frac{c}{2}\right)^2 + \left(\frac{b}{3}\right)^3}} - \sqrt[3]{-\frac{c}{2} + \sqrt{\left(\frac{c}{2}\right)^2 + \left(\frac{b}{3}\right)^3}}.$$

Of course, Cardano never wrote such a formula. In his times there were no formulas at all. On the surface algebra was still *regula della cosa*, a system of verbal rules used to find the thing.

2.2.2. The Projective Form of the Language of Algebra

In the previous section I presented Cardano's rule for solution of the cubic equation. Even if the rule itself did not deviate from the framework of al-Khwárizmí's approach to algebra, it is not clear how it was possible to discover something so complicated as this rule. In order to understand this, we have to go back a century before Cardano and describe the first stage of the reification of the language of algebra, the stage connected with the creation of algebraic symbolism. After western civilization had absorbed Arabic algebra, a tendency arose to turn algebraic operations into symbols and in this way to make them visible. This process was slow, lasting nearly two centuries, and most likely those responsible for it were not fully aware of it. We will present only some of the most important innovations. Regiomontanus introduced the symbolic representation for root extraction. He denoted the operation of root extraction with the capital R, stemming from the Latin *radix*. Thus for instance he expressed the third root of eight in the form *R cubica de 8*. In this way he represented the *operation* of root extraction by the *expression* of the root itself, that is by the result of the operation. Stifel replaced the capital R by a small r, so that instead of *R cubica de 8* he wrote $\sqrt{c}8$. He introduced the convention to write the upper bar of the letter r a bit longer. Stifel placed the first letter of the word *cubica* below this prolonged bar, so that everybody would know that it was the cube root. The number placed after this sign is the one whose root is to be extracted. This convention is quite similar to that used by Regiomontanus, differing only in that it is a bit shorter, using only the first letter of *cubica* instead of the whole word. Nevertheless, this small change opened the door to our modern convention, which was introduced by Descartes. Descartes replaced Stifel's letter c by the upper index, and placed the number itself below the bar of the letter r, thus writing the third root of eight as $\sqrt[3]{8}$.

Another very important development took place in connection with representation of the unknown. The Arabic terms of *shai*, *mál*, and *kab* were translated as *res*, *zensus*, and *cubus*. Perhaps because the latter two are rather long, instead of writing the whole words mathematicians started to use only their first letters, thus r for *res*, z for *zensus*, and c for *cubus*. Just like the Arab algebraists, the *Cossists* (as the practitioners of this new algebra were called) did not stop with the third power of the unknown, introducing higher powers, such as zz (*zenso di zensi*), zc (*zenso di cubo*), etc., and developing simple rules for calculating with such expressions. Through such gradual processes symbols for the algebraic operations were introduced and an algebraic symbolism appeared. The operation of addition was represented by the symbol $+$, the operation of root extraction by $\sqrt{\ }$, and gradually a whole layer of operations was reified, *acts were turned into objects*. This process was slow, and at the beginning it was only little more than replacing of words by letters for the sake of brevity. Nevertheless, when the new symbols accumulated in sufficient quantity, they made possible a radical change in algebraic thought – the solution of the cubic equation. Cardano formulated his result in traditional form, as a verbal rule. Nevertheless, its discovery was made possible by the new symbolism. We will present a reconstruction of this discovery, using modern symbolism for the sake of comprehensibility (see van der Waerden 1985 pp. 53–56, or Scholz 1990 pp. 165–172).

Let us take a cubic equation $x^3 + bx = c$, i.e., Cardano's "*De cubo et rebus equalibus numero*".[18] The decisive step in the solution of this equation is the assumption that the result will have *the form of* the difference of two cube roots. When we make this assumption, everything becomes simple. Let

$$x = \sqrt[3]{u} - \sqrt[3]{v}.$$ (2.5)

Raising this expression to the third power and then comparing it with the equation we obtain the following relations between the unknown

---

[18] The other two types of cubic equations $x^3 = bx + c$ (*cubus is equal to things and number*) and $x^3 + c = bx$, (*cubus and number is equal to things*) can be solved in an analogous way. Cardano did not consider equations of the kind $x^3 + bx + c = 0$ because he considered all numbers entering an equation to be positive and so their sum cannot be zero.

quantities u and v, and the coefficients b and c:

$$b = 3\sqrt[3]{uv}, \qquad\qquad c = u - v. \qquad (2.6)$$

When we isolate v from the second equation, and substitute it into the first one, we obtain the equation

$$u^2 - uc - \left(\frac{b}{3}\right)^3 = 0. \qquad (2.7)$$

The root of this equation can be found by the help of the standard formula for quadratic equations as

$$u = \frac{c}{2} + \sqrt{\left(\frac{c}{2}\right)^2 + \left(\frac{b}{3}\right)^3}. \qquad (2.8)$$

The value of the unknown v can now be determined from the second equation of (2.6). Knowing u and v we can find the solution of the original problem from (2.5) in the form

$$x = \sqrt[3]{u} - \sqrt[3]{v}$$

$$= \sqrt[3]{\frac{c}{2} + \sqrt{\left(\frac{c}{2}\right)^2 + \left(\frac{b}{3}\right)^3}} - \sqrt[3]{-\frac{c}{2} + \sqrt{\left(\frac{c}{2}\right)^2 + \left(\frac{b}{3}\right)^3}}.$$

In this derivation the advantage of the algebraic symbolism, i.e., of the reification of the language of the Arabic algebra, is clearly visible. Right at the beginning we assumed that the result would have *the form of* the difference of two cubic roots. In Arabic algebra there were no expressions, there were only rules. And *a rule has no form*, because it cannot be perceived. We can only listen to it, and then perform all the steps precisely as the rule instructs us. Only when we represent the steps of the rule by symbols does the sequence of calculations appear before our eyes; only then are we able to perceive its form. *The rule is thus transformed into a formula.*

Step (2.7) is also noteworthy. We obtained an equation for u. But what is this u? It is not a thing determined by the original equation. It does not stand for anything real. If we were to compare the equation to a picture we would see that here we have a *representation of a representation*, a picture inside a picture. From the technical point of view it is the decisive step in the whole process of solution, because instead of a cubic equation for the original unknown x, we obtain a quadratic equation for this auxiliary unknown, which we already know how to solve. From the epistemological point of view it is a fundamental shift. In the context of the original equation the unknown x has a real denotation: we know what it represents, what it stands for. But now it is replaced by some u, about which we know nothing. Its meaning is determined only through the equation (2.5). Thus the unknown u has no direct reference. Its *reference is given only indirectly*, through the reference of x.

We suggest calling this layer of the language of algebra the *projective form*. The paradigmatic example of the projective form of language in geometry was the drawing of Dürer that introduced a similar loss of directness of reference. In Dürer's drawing there is a picture of a vase inside of the picture, while in algebra we have the equation (2.5) relating an unknown u to another unknown x. In this way algebra stops being *regula della cosa* and it becomes an *analytic art*, the art of transforming algebraic formulas, the art of guessing the form of the result, the art of finding suitable substitutions. This art resembles projections, which form the core of projective geometry. In both cases we have to deal with transformations of linguistic expressions that preserve reference.

Nevertheless, it is important to realize that new possibilities are opened up when one reifies one level of the algebraic language by turning the rules into formulas (i.e., when the rule "square the thing and add to it five things" becomes simply "$x^2 + 5x$"). Al-Khwárizmí knew three algebraic operations – *al-jabr, al-muqábala*, and *al-radd* – but not *substitution*. His transformations did not make it possible to change the "form" of the algebraic formulas. In this respect al-Khwárizmí's algebra resembles Euclidean geometry, which also studied only form-preserving transformations of geometric figures (parallel translations, rotations, and uniform changes of scale). Central projections, which can change the form of a figure (e.g., it can transform a circle into a hyperbola), were studied in projective geometry. I would like to stress a fundamental analogy between central projections and substitutions. On the basis of this analogy, I suggest that we call this stage in the development of algebra, which is characterized by the emergence of substitutions, as the stage of the *projective form of the language of algebra*. A substitution does not simply shift a term as a whole from one side of the equation to the other but rather decomposes it and then rearranges its parts in a new way. For instance, the substitution $x = \sqrt[3]{u} - \sqrt[3]{v}$

decomposed the unknown x into two parts, rearranged these parts, and then put them together again. Thus it seems that the transition to the projective form *shifts the ontological foundations* one level deeper. Algebra as *regula della cosa* understood the unknown as a "thing", that we can "take in our hands and move to some other place", but the "form" of this thing remained unchanged. In the projective form of language the "thing" itself is transformed. It is, for instance, decomposed into two parts which can be treated separately. Thus the projective form of the language of algebra can be characterized by *indirect reference* (auxiliary variables), *representation of representation* (auxiliary equations), and *transformations of expressions changing their form* (substitutions). These three aspects are interdependent; each of them presumes the others.[19]

Another important change introduced by the use of substitutions concerns what counts as the solution of an algebraic equation. Formerly, in the framework of algebra understood as *regula della cosa*, mathematicians accepted only positive solutions, because the number of the things (*cosa*) cannot be negative. If the unknown represents some real quantity, some number of real things, it cannot be less than nothing. But as soon as we start using auxiliary equations, the unknowns of which have only indirect reference to reality due to substitutions, it can happen that the positive solution of the original equation corresponds to a negative root of the auxiliary equation. Therefore at least in the auxiliary equations we have to take into account *the negative as well as the positive roots*. In this context Cardano distinguished between the "false" and the "true" solutions. The notion of an equation is slowly freed from its dependence on direct reference. To add the "false" to the "true" solutions was necessary in order to turn the transformations of the formulas into equivalent transformations. Thus as a further aspect of the projective form of language we can mention the *extension of the language* by expressions with no direct denotation (infinitely re-

---

[19] The fundamental changes that appeared in science at the break of the fifteenth and sixteenth centuries were in many respects analogous. The discovery of Copernicus that the Earth is moving is based on the ability to look at the Earth so to speak from the outside, from a viewpoint situated somewhere outside of our solar system. From this external point of view we can see that, what from the internal viewpoint, situated on the Earth, appears as the diurnal rotation of the skies is in reality (i.e., from the external viewpoint) caused by the Earth's rotation. On a similar doubling of viewpoints is based also the discovery of the solution of the cubic equations. A substitution means in a sense to take an external viewpoint. And the observation of the process of painting from an external viewpoint is represented also in Dürer's drawing.

mote points, negative roots), which make smooth functioning of the language possible.

The most important shortcoming of the previous algebraic notation was that it used different letters ($r, z, c, \ldots$) to represent the powers of the same quantity. Thus, for instance, if r is 7, then z must be 49, but this dependence is not indicated by the symbolism. When substitutions are used, such a convention becomes unwieldy, because whenever we make a substitution for r, we must also make the appropriate substitution for z. Further, in a substitution we have to deal with at least two unknowns, the old one and the new one. To represent both with the same letter r would create an ambiguity. Another shortcoming of the symbolism of the *Cosists* was that it had no symbols for the coefficients of the equation. Instead they used such phrases as "the number of things", meaning the coefficient of the first power of the unknown. They were unable to express the coefficients in a general way.

In 1591 Viète published his *Introduction to the analytic art*. In this book Viète introduced the symbolical distinction between unknowns and parameters. He was the first to represent the coefficients of equations with letters. Viète used capital vowels A, E, I, O, U, to represent the *unknowns* and the capital consonants $B, C, D, F, G, \ldots$ to represent the *coefficients*. In addition, each quantity had a dimension: 1-*longitudo*, 2-*planum*, 3-*solidum*, 4-*plano-planum*, $\ldots$. The dimension of each quantity was expressed by a word written after the symbol, thus for instance *A-planum* was the second power of the unknown A while *A-solidum* was the third power *of the same unknown*. The letter indicates the identity of the quantity while the word indicates its particular power. This makes it possible to use more than one quantity, and thus to express a substitution. Viète understood quantities as dimensional objects and therefore retained the principle of homogeneity. This principle stipulated that we can add or subtract only quantities of the same dimension. This is a carryover from geometry, because in geometry we cannot, for example, add a length to a volume. Even though Viète's symbolism is rather complicated, it was *the first universal symbolic language for the manipulation of formulas*. Viète was fully conscious of the importance of his discovery. He believed that this new universal method, or analytic art, as he called it, would make it possible to solve all problems. Viète ended his book with the words: "*Finally, the analytical art, having at last been put into the threefold form of zetetic, poristic, and exegetic, appropriates to itself by right the proud problem of problems, which is: TO LEAVE NO PROBLEM UNSOLVED.*"

2.2.3. The Coordinative Form of the Language of Algebra

One of Cardano's merits was the systematic nature of his work. There-fore besides the equation of the form "*cubus and thing equal number*", the solution of which was discussed above, he presented rules for solu-tion of the other two forms of cubic equations. The rules for solution of these equations have a form very similar to the first case, and can be obtained by simple substitutions. Therefore I will not discuss them here.[20] Instead, let us turn to a discovery Cardano made when he tried to apply his general rule for the equations of the form "*cubus equals thing and number*" to the equation $x^3 = 7x + 6$. When he applied the rule he obtained a result which we would express as:

$$x = \sqrt[3]{3 + \sqrt{-\frac{100}{27}}} + \sqrt[3]{3 - \sqrt{-\frac{100}{27}}}.$$

Below the sign for the square root a negative number appeared. The formula required him to find $\sqrt{-\frac{100}{27}}$, something he was not able to do. Cardano discovered something from which *complex numbers*[21] evolved. For the further progress of algebra it was crucial to under-stand what was going on when a negative number appeared below the square root sign. The rules of manipulation with formulas brought Car-dano into a situation that was beyond his comprehension, a situation where he had to do something impossible. From the point of view of the projective form of language, the square root of a negative number is a meaningless expression.

The way out of this situation led to a change of our attitude towards formulas. In the projective form, even if the reference of the language becomes indirect, the linguistic expressions are still understood as tools for representing some real objects. Nevertheless, the discovery of the *casus irreducibilis*, of the insoluble case, led to a gradual loosening

---

[20] The result $x = \sqrt[3]{\frac{c}{2} + \sqrt{\left(\frac{c}{2}\right)^2 - \left(\frac{b}{3}\right)^3}} + \sqrt[3]{-\frac{c}{2} + \sqrt{\left(\frac{c}{2}\right)^2 - \left(\frac{b}{3}\right)^3}}$ is the solution of the equation $x^3 = bx + c$. It may seem that only small changes took place – beneath the sign of the square root addition was replaced by subtraction. But these minute changes have radical consequences.

[21] Complex numbers are numbers of the form $a + b\sqrt{-1}$, thus for instance $3 + 7\sqrt{-1}$. They are called complex because they are composed of two parts. One part of them is real (the number 3 in our case) and the other imaginary (the number $7\sqrt{-1}$). $\sqrt{-1}$ is called the imaginary unit and is often represented by the letter i. We usually write $2 + 3i$ instead of $2 + 3\sqrt{-1}$.

of the bond between language and reality. Algebraic expressions are viewed more and more as *forms*,[22] as formal objects constructed from symbols, independent of any realistic context in which they are supposed to be interpreted. An important motive for a development in this direction was the situation in the theory of equations. Cardano considered equations such as $x^3 + bx = c$ and $x^3 = bx + c$ to be different problems, because he allowed only positive numbers for coefficients and solutions. For equations of the third degree this represents only a small complication, but in the case of the equation of fourth degree we have seven different kinds of equations, and in the case of quintic equations fifteen. These kinds of equations are not fully independent, because, as Cardano had shown, simple substitutions can transform an equation of one kind into another. Therefore it is rather natural to try to reduce this complexity. It was Michael Stifel, whom we have already mentioned in connection with the introduction of the symbol for the square root, who first saw how this might be accomplished. In his book *Arithmetica integra* (1544) he introduced rules for the arithmetic of negative numbers, which he interpreted as numbers smaller than zero. That is an extension of the number concept, natural for the projective form of language, because negative numbers begin to play an important role as values of the auxiliary variables. But Stifel went further, beyond the realm of the projective form of language, and started to use negative numbers also as coefficients of equations. This enabled him to unite all fifteen kinds of quintic equations, which formerly had to be treated separately, into one general *form*: $x^5 + ax^4 + bx^3 + cx^2 + dx + e = 0$.

Before Stifel attempted to solve an equation, he transferred all the expressions to one side, and in this way obtained the equation in a *polynomial form* $p(x) = 0$. Thus a polynomial as a mathematical object was first found in Stifel's work. He used a simple symbolism without symbols for coefficients. But the basic idea of reducing all the different cases to a single form, by allowing the coefficients to be negative, was decisive. A polynomial is an expression that co-ordinates different *formulas* into one universal *form*. Accordingly, we call the form of language of algebra based on this approach the *coordinative form*. The form common to all cubic equations had remained hidden until then,

---

[22] The term *form* is used here in two different ways. On the one hand the term form is used in the sense *"form of the language of algebra"* and so it enters such phrases as *"the perspectivist form"*, or *the "projective form"*. In this sense it is a term belonging to epistemology. On the other hand we use the term form in the algebraic sense in phrases such as *"polynomial form"*.

because mathematicians bound the algebraic language too firmly to reality. Only when they stopped observing a distinction between positive and negative coefficients did this new deeper unity become visible.

When we free ourselves from the understanding of the algebraic expressions as pictures of reality, and start to understand them as independent formal objects, it becomes possible to accept the square roots of negative numbers simply as a special kind of expressions. Even though we do not know what such expressions represent, we know how to calculate with them. This understanding is implicit in the *Algebra*, written by Bombelli in 1572. Bombelli simply introduced rules for the addition, subtraction, multiplication and division of these new expressions, and did not ask what they stood for. Maybe the most pregnant expression of this view can be found in Euler's book *Vollständige Anleitung zur Algebra* from 1770, where imaginary numbers are called *numeri impossibiles*, because they are not smaller than zero, not equal to zero, and not greater than zero. Euler writes:

> "But notwithstanding this, these numbers present themselves to the mind; they exist in our imagination, and we still have a sufficient idea of them; since we know that by $\sqrt{-4}$ is meant a number which, multiplied by itself, produces -4; for this reason also nothing prevents us from making use of these imaginary numbers and employing them in calculations.." (Euler 1770, § 145)

Thus even though in reality there is no quantity whose square is negative, we have a clear understanding of the meaning of the symbol $\sqrt{-4}$.

The transition from formulas to forms is also important for another reason. If we consider formulas as the basic objects of algebra, one central aspect remains hidden. A polynomial has many roots; thus in general there is not just one number satisfying the conditions of an algebraic problem. In the framework of the *perspectivist form of language* this was ignored, because in reality, in the normal case, the number of things we are looking for is unequivocally determined. Mathematicians therefore ignored the existence of other roots of an equation, and as the solution of the problem they accepted the root that made sense given the problem's context. In most cases they were not even aware that they were overlooking some solutions, because in most cases the other solutions were negative, and thus – from the point of view of the perspectivist form of language – unacceptable. In connection with the *projective form of language* the situation was somewhat better. For the auxiliary equations it was necessary to take the negative solutions into account as well, because it can happen that a negative solution of the auxiliary equation corresponds to a positive solution of the original problem. But as a solution of the whole problem mathematicians still accepted only a positive number, one that gave the "number of things". Only when the bonds tying the language to reality became looser did they accept that equations generally have more roots. Thus the transition from formulas to forms was crucial for understanding of the relation between the degree of an equation and the number of its roots.

We expect a *formula* to tell us the result, to give an answer to the problem. A formula expresses a number we want to know, it represents the answer to the question we are asking. A *form*, on the other hand, is a function, giving different results for different inputs. It might not be easy to imagine that a given problem has more than one answer, because if we are asking something about reality, we expect that the answer is uniquely determined. Yet when we understand the equation describing the problem as a polynomial form, it becomes understandable that the form can produce the same value (usually zero) for more than one value of its argument. Thus the transition from formulas to forms makes it easier to accept that an equation can have more than one solution. Already Viète knew about this, because he had discovered the relation between the roots and the coefficients of an equation. But oddly enough all the roots he presented in his examples were positive. Only when the equation is understood as a form can the relation between the roots and the coefficients be disclosed in its entire generality, as was done independently by Girard and Descartes.

The notion of a form attained its full meaning in the work of Descartes. From Euclid to Descartes a product of two quantities was understood as a quantity of a new kind. Thus a product of two segments was a rectangle. The product of a rectangle with a segment gave a prism. It is true that the *Cosists* went beyond the three dimensions allowed in geometry, but the terms they used for the powers of the unknown show that they still thought of algebraic operations in geometrical terms. The influence of geometry is even more visible in Viète. Descartes left this tradition behind when he introduced a radically new interpretation of algebraic operations. For him for the first time the product of the segments x and y is not a rectangle, having an area equal to xy, but a segment having the length xy. This change was one of the most important ideas in the development of algebra. When Descartes interpreted the product of two segments again as a segment, he created a system of quantities closed under algebraic operations. Thus with a slight touch of anachronism we can say that *Descartes created the first example of a field*. By a field we understand a system of quantities containing 0 and 1 that is closed under the four algebraic operations. From the epistemological point of view closure under operations means the grasping of the universe of discourse as a whole.

Thus algebraic language served a fundamentally new function, the function of grasping the unity of the world represented by the language, grasping the way its different aspects are *co-ordinated*. This co-ordination exists on two levels. On the one hand we have the co-ordination of different formulas (for instance the different kinds of equations as *cubus and things equal number* and *cubus equals things and number*) into a single form of a polynomial. On the other hand we have a co-ordination of different sorts of quantities (for instance Viète's *longitudo, planum, solidum, plano-planum*, etc.) into a single field. Thus the language of algebra becomes a means for grasping the unity behind the particular formulas and quantities. This unity opens up a new view of equations. Instead of searching for a formula that would give us the value of the unknown, we face the task of finding all the numbers that satisfy the given form. We are searching for numbers which we can use to split the form into a product of linear factors, as for instance

$$x^3 - 8x^2 + x + 42 = (x - 7)(x - 3)(x + 2),$$

which shows that 7, 3 and -2 are the roots of the polynomial $x^3 - 8x^2 + x + 42$. When we have found all the roots, we are able to split the form $x^3 - 8x^2 + x + 42$ into linear factors $(x - 7)$, $(x - 3)$ and $(x + 2)$. This factorization shows that no other root can exist (for any number different from 7, 3 and -2 each factor gives a nonzero value and so their product is also nonzero). Thus the splitting of the form into linear factors gives a complete answer to the problem of solving a given equation. From the point of view of the co-ordinative form, to solve an equation means to *split a form into its linear factors*.

2.2.4. The Compositive Form of the Language of Algebra

After Girard and Descartes discovered that a polynomial of degree n has precisely n roots, Cardano's "formulas" had to be revised. These formulas determined only one root of the cubic equation. But a cubic equation has three roots, and so it became necessary to find a way to determine the remaining two roots. Johann Hudde found a procedure that made it possible to find all the roots of a cubic equation. Consider the equation $x^3 + px - q = 0$. Using the substitution $x = y - \frac{p}{3y}$ Hudde transformed this equation into

$$y^6 - qy^3 - (p/3)^3 = 0,$$

which was later named *Hudde's resolvent*. Despite the fact that it is a sixth-degree equation, it is simpler than the original equation, because by the substitution $y^3 = V$, we obtain a quadratic equation

$$V^2 - qV - (p/3)^3 = 0,$$

which has the roots

$$V_1 = \frac{q}{2} + \sqrt{\frac{q^2}{4} + \frac{p^3}{27}} \quad \text{and} \quad V_2 = \frac{q}{2} - \sqrt{\frac{q^2}{4} + \frac{p^3}{27}}.$$

In order to get from the variable V back to y, we could take the cube root of V. Nevertheless, that would be a mistake, because in this way we would obtain only two roots of the original equation, one as the cube root of V_1 and the other as the cube root of V_2. But y is the root of an equation of the sixth degree, which has six roots. Hudde realized that the operation of root extraction is the step where we are losing roots. If in the equation $y^3 = V$ we substitute a number for the variable V, for instance 1, the equation $y^3 = 1$, as an equation of the third degree, has three roots. Thus besides the root $y = 1$, there are two others. These are the complex cube roots of unity. Their values are

$$\omega = -\frac{1}{2} + i\frac{\sqrt{3}}{2} \quad \text{and} \quad \omega^2 = -\frac{1}{2} - i\frac{\sqrt{3}}{2}.$$

Here we used i for $\sqrt{-1}$. With the help of ω and ω^2 we can find all roots of Hudde's resolvent:

$$y_1 = \sqrt[3]{V_1}, \qquad y_2 = \omega \cdot \sqrt[3]{V_1}, \qquad y_3 = \omega^2 \cdot \sqrt[3]{V_1},$$
$$y_4 = \sqrt[3]{V_2}, \qquad y_5 = \omega \cdot \sqrt[3]{V_2}, \qquad y_6 = \omega^2 \cdot \sqrt[3]{V_2}.$$

The roots of the third-order equation, from which we started, will then be combinations of these roots:

$$x_1 = y_1 + y_4 = \sqrt[3]{V_1} + \sqrt[3]{V_2},$$
$$x_2 = y_3 + y_5 = \omega^2 \cdot \sqrt[3]{V_1} + \omega \cdot \sqrt[3]{V_2},$$
$$x_3 = y_2 + y_6 = \omega \cdot \sqrt[3]{V_1} + \omega^2 \cdot \sqrt[3]{V_2}.$$

We obtained each of the three solutions of the cubic equation in the form of a combination of two cubic roots, that is, in the same form as in Cardano's formulas. Nevertheless, the whole procedure is much more systematic.

Euler tried to generalize Hudde's approach for the case of a general polynomial $f(x) = x^n + a_1 x^{n-1} + \ldots + a_n = 0$. He wanted to find an analogous auxiliary equation, the so-called *Euler's resolvent*, the roots y_i of which would be connected to the roots of the original polynomial in a way similar to that in which the roots of Hudde's resolvent are connected with the roots of the original cubic equation. But Euler did not get very far, because the degrees of the resolvents he obtained were too high. For the cubic equation the degree of the resolvent was 6, for the biquadratic the degree of the resolvent was 24, and for the quintic, which was the main point of interest, the degree was 120. Euler could not find any trick by means of which it would be possible to reduce the degree of the resolvent.

At that point Lagrange entered the discussion with a generalization of Euler's approach. Lagrange introduced a new kind of resolvent, which today bears his name. The *Lagrange resolvent* for a polynomial of the fourth degree does not have the degree 24, as Euler's resolvent did, but only the degree 3. Thus Lagrange's resolvent seemed to be a step forward, because it made it possible to reduce the problem of solving an equation of the fourth degree to the solving of its resolvent, which was of the third degree. Lagrange succeeded in showing that all tricks for solving equations of the third and the fourth degree, presented in Cardano's *Ars Magna*, were based on the use of resolvents. It is sufficient to look at the solution of the cubic equation, where an auxiliary quadratic equation (2.7) appeared. The difference was that, while Cardano found this auxiliary equation by chance, in Lagrange's approach the resolvent appears in a systematic manner, and its importance is fully understood. Thus what was previously only a piece of good luck becomes a consciously applied method with Lagrange. All the techniques for solving algebraic equations that had been discovered before Lagrange consisted explicitly or implicitly in the reduction of the initial equation to its resolvent, which was of lower degree. After gaining this insight into the process of solution of algebraic equations, Lagrange believed that the technique of reduction of an equation to its resolvent should also work for quintic equations, where the resolvent should be of the fourth degree. But here Lagrange met with a disappointment: the resolvent of the equation of the fifth degree turned out to be of the sixth degree. Six is much less than 120, the degree of Euler's resolvent, but it is still more than five, and so this resolvent is of no help in solving the original equation. Lagrange was thus confronted with a strange situation. His method worked nicely for equations of the third and fourth degree, where the resolvents had lower degrees than the initial equations, but in order to solve the equation of the fifth degree, his method required the solution of an equation of the sixth degree.

But even if the methods of Hudde, Euler, and Lagrange did not work for quintic equations, they do have something important in common. All three of them are trying in some way or other to *compose* the solution x of the original problem from the solutions y of the resolvent and the complex roots ω of the unit. Of course with the advantage of hindsight we see here the first step towards the notion of factorization. But leaving this aside for the moment, we can say that the *compositive form of language* is based on the idea of decomposing a problem into independent parts that are easier to handle. Thus instead of alternative perspectives on the same problem, which was the basis of the projective form of language, in the compositive form we have to do with a variety of analogous problems arranged in such a way that the solution of the more complicated problem can be reduced to the solution of the preceding, simpler problems.

2.2.5. The Interpretative Form of the Language of Algebra

The interpretative form of language is connected with acceptance of the square roots of negative numbers as standard mathematical quantities. The acceptance of these so-called complex numbers amounts to a reification of another layer of the language of algebra. Nevertheless, this reification was very different from the previous one connected with the birth of algebraic symbolism. In the case of algebraic symbolism, specific acts were reified. For instance, the operation of taking the cube root of a number was turned into a symbol representing the result of this operation, or the operation of addition was turned into a symbolic representation of the sum (i.e., of the result of this addition). By contrast, in the case of the interpretative form, the reification does not consist in turning the results of some additional operations into objects. It is impossible to take the square root of a negative number; thus, strictly speaking, there is nothing to be reified. We have at our disposal no concrete activity, no performance, the result of which we could declare to be a new object. A square root of a negative number instead represents the impossibility of an operation, a failure of the language. The process that was brought to its completion by construction of the model of the complex plane therefore consisted in something fundamentally different from simply turning operations into objects.

This process started with overcoming the antipathy towards expressions containing square roots of negative numbers and simply adding them to the language of algebra. The square roots of negative numbers were considered to be a special kind of expression; while their use was well understood, it was not clear what they signified. Already in the perspectivist form of language of algebra new expressions appeared, the negative numbers, which also had no direct reference. Thus it seemed reasonable to consider the complex numbers as something similar to the negative numbers, that is, expressions of the language whose reference is indirect. Nevertheless, such an interpretation of the complex numbers is doomed to failure. A negative number can be turned into a positive one by means of an appropriate substitution, and so may be regarded as a representation of a representation. In the case of a complex number nothing like this is possible. A complex number can never indicate any number of things, directly or indirectly, because the complex numbers cannot be linearly ordered, or in Euler's words *"they are neither lesser nor greater then nothing"*.

Another attempt at an interpretation of the square roots of negative numbers was based on the idea of ascribing them not an indirect but rather a subjective meaning. Descartes introduced the term *imaginary number* for the square root of a negative number[23]. In his *Geometry* he writes that in the case of an equation that does not have enough true (*vraies*, i.e., positive) and false (*fausses*, i.e., negative) roots, we can imagine (*imaginer*) some further, imaginary (*imaginaires*) roots. Nothing in reality corresponds to the imaginary roots. Descartes introduces them only in order to ensure that an equation of the *n*th degree

---

[23] From this phrase of Descartes the symbol i for $\sqrt{-1}$ was created.

will have *n* roots. Euler has this to say about the meaning of the square roots of negative numbers:

> "they are neither nothing, nor greater than nothing, nor less than nothing; which necessarily constitutes them imaginary, or impossible. But notwithstanding this, these numbers present themselves to the mind (Verstand); they exist in our imagination (Einbildung), and we still have a sufficient idea of them.." (Euler 1770, §§ 144–145)

Thus for Euler too these quantities exist only in our imagination. But this subjective interpretation of the complex numbers cannot explain how it is possible for computations involving these non-existent quantities to lead to valid results about the real world. It is as if a biologist, after reflecting on centaurs, were able to bring forth new knowledge about horses, and have his claims substantiated by biology. If the complex numbers make it possible to disclose new knowledge about the world, they must be related to the real world in some way. A purely subjective interpretation is therefore unsatisfactory.

In 1799 Gauss created a geometric model of the complex numbers. The problems with the interpretation of the complex numbers were not solved by finding how to ascribe to an individual complex number some reference in the context of the particular problem, in the solution of which that complex number appeared. Instead, all possible expressions to which we cannot ascribe a reference were reified. Rather than seeking interpretations of individual complex numbers in particular contexts, a model is constructed for all of them at once. Gauss's complex plane was thus perhaps the *first model in the history of mathematics*. For the first time an artificial universe of objects was constructed in which the language as a whole is interpreted. Thus the idea of a model appeared in Gauss's work some 70 years before Beltrami. From the epistemological point of view, Gauss's model of complex numbers is similar to Beltrami's model of non-Euclidean geometry. First of all, in both cases the model serves the purpose of making a doubtful theory acceptable. Before Gauss the complex numbers had a dubious status, just as before Beltrami the status of the non-Euclidean geometry was not clear. The fundamental similarity between Gauss's plane and Beltrami's model lies in that both of them *actualize a whole world*. When Beltrami constructed a model of the non-Euclidean plane inside a circle, he actualized the whole world of non-Euclidean geometry. In a similar way, Gauss's plane represents the whole world of complex numbers. At first sight it might seem that some regions must escape when he represents the complex numbers by means of a plane, because the plane he uses cannot be viewed as a whole. But we must remember that Gauss's model concerns not geometrical lines but rather algebraic expressions. Therefore even if from the geometrical point of view Gauss's plane escapes our field of vision, from the algebraic point of view it represents the whole world – because the world of algebra is a world of operations. Therefore to represent the whole world of algebra does not mean that it is *"entirely in the field of vision"* (as it is in geometry), but to be *"closed under operations"*. And Gauss showed that the sum, difference, product, and quotient of two points of the complex plane is again a point of this plane. Therefore Gauss's plane is a closed universe, in which we can interpret all algebraic operations. A third similarity between Gauss's and Beltrami's models consists in the fact that in both of them a *translation is incorporated into the language*. Gauss ascribes to each complex number a point of the plane, and he shows how it is possible to translate the algebraic operations of addition or multiplication of complex numbers (i.e., operations of the internal language of the model) into the language of geometric manipulations with the points of the plane (the external language of the model).[24] On the other hand Beltrami ascribes to every figure of the non-Euclidean plane a figure inside a circle of the Euclidean plane, and he shows how it is possible to translate the notions of non-Euclidean geometry (i.e., expressions of the internal language of the model) into the language of Euclidean geometry (the external language of the model).

With the help of his model Gauss proved the *fundamental theorem of algebra*, which says that every polynomial of the nth degree has n roots. In proving this theorem Gauss showed that the problem with the quintic equations, for instance the equation $x^5 - 6x + 3 = 0$, does not lie in the fact that they lack roots. In the complex plane there are exactly five points corresponding to roots of a quintic equation – i.e., points whose co-ordinates, when substituted into the equation, return the value zero. Thus the problem with the quintic equations turns out to be more subtle. Even though the roots of such equations exist, *it is impossible to express them by means of the language of algebra*. That is, there is no general formula formed from integers, the four arithmetic operations $(+, -, \times, \div)$ and root extraction ($\sqrt[5]{}$, $\sqrt[17]{}$, $\sqrt[542]{}$, ...), which represents the roots of the above equation.

---

[24] For instance the complex number $7 + 2i$ can be represented by a point of the complex plane having the x co-ordinate 7 and the y co-ordinate 2.

To see more clearly what the problem is, let us imagine a huge sheet of paper, on which all the algebraic formulas are already written. Thus on our paper we have formulas consisting of 500, 1 000, 1 000 000 or any other number of symbols. We would like to prove that none of the five roots of the equation $x^5 - 6x + 3 = 0$ is to be found on this piece of paper. How can we prove such a claim? It is easy to prove that when we add, subtract, multiply or divide any two algebraic formulas (excepting division by 0), we again obtain an algebraic formula. This means that all the formulas contained on our paper form a closed system, which is in algebra called a *field*. We want to show that this field does not contain any root of the above-mentioned polynomial. Here we see the advantage the interpretative form of language yields. By reifying the whole world of algebraic expressions a *modal predicate*, that something it is impossible to do (an equation cannot be solved) is turned into an *extensional predicate*, that some numbers do not belong to a field. Due to this reification of the world of algebraic expressions we begin to see what the problem might be with quintic equations. The problem is not that these equations lack solutions, but rather that the language of algebra is not rich enough to express these solutions. Thus in a sense the reification of the two worlds, Gauss's reification of the world of complex numbers and the above-described reification of the world of algebraic formulas, make it possible to understand the phenomenon of insolubility. The insolubility of the quintic equations means that their roots (which, as complex numbers do exist) fall outside the world of algebraic formulas. Thus the language of algebraic formulas and the world of roots of algebraic equations do not fit together. But even if the interpretative form of language were able to understand what the problem with quintic equations might be, it does not have the means to prove that the general quintic is insoluble. To be sure, some simpler problems – simpler, that is, from the point of view of the fields involved – like the impossibility of trisecting an angle, could be handled at this stage. But in order to prove that quintic equations are insoluble, it was necessary to reify the next layer of the language of algebra and in this way to create a much stronger tool – group theory.

2.2.6.   The Integrative Form of the Language of Algebra

In the framework of the interpretative form of the language of algebra, the whole world of algebraic formulas was reified. The world of algebraic formulas is a field, that is, a system of objects closed under the four basic arithmetical operations. Nevertheless, closer investigation reveals that inside this world there is a whole range of different sub-fields, a whole range of smaller worlds, from the smallest one, the field Q of all rational numbers, to some slightly bigger fields like $Q(\sqrt{2})$ to the greatest field of all, the complex numbers C. Gauss showed that in the field of all complex numbers each polynomial of the degree n has n roots, and so the polynomial $x^5 - 6x + 3$ has five roots. The field of numbers that can be expressed by algebraic formulas lies somewhere between the fields Q and C. It is richer than the field Q of all ratio-nal numbers because it contains irrational numbers such as $\sqrt{2}$. On the other hand it is poorer than the field C of all complex numbers, because the number π is not expressible by any algebraic formula. In order to show that the roots of the equation $x^5 - 6x + 3 = 0$ can-not be expressed by any algebraic formula, we have to characterize which numbers can be expressed by such formulas. But the interpreta-tive form of language is not up to this task. That form is able to reify a whole world and turn it into an object (a field), and to describe the transition from one such world to another (a translation). But it can-not compare different worlds – because it is always restricted to one abstract structure. Thus, for instance, in the construction of the Gaus-sian model of the complex numbers we are able to translate algebraic operations with complex numbers into geometrical transformations of points of the Gaussian plane, but we are able to do this only because the structures of the complex numbers and the Gaussian plane are isomor-phic. In a way, therefore, we are always working with the same abstract structure, and move only between its different realizations. The formal relations are the same in both cases and so we can say that the interpre-tative form is able to reify only one structure, and move this structure from one medium into another.

Only the transition to the integrative form of language, which re-places the translation between equivalent structures (isomorphism) by an embedding into a richer structure (homomorphism), makes it pos-sible to compare different structures. Again there is a whole range of analogies between how Cayley and Klein introduced the integra-tive form of language in geometry and how Galois introduced the in-tegrative form of language in algebra. The first common feature is the existence of a *neutral basis*, a fundamental level of description in terms of which all the structures are to be compared. In geometry the projective plane was such a basis, and the different geometries were compared as structures introduced into this neutral basis. In algebra the field of complex numbers is such a neutral basis, and all the fields Galois was working with are subfields of this field. The next common feature is the role played in both cases by *group theory*. In geometry, Klein compared the different geometries by comparing the transformations groups associated with them. In algebra, group theory is the basic means which makes it possible to compare the different fields. Thus it can be said that the integrative form of language integrates different (geometric or algebraic) worlds by embedding the symmetries of these worlds (transformation groups or groups of automorphisms) into one neutral structure (the projective plane or the field of all complex numbers). Nevertheless, the proof of the insolubility of the quintic equation is a bit too complicated from the technical point of view and I have therefore decided to split it up into its basic steps in order to make it more comprehensible.

2.2.6.1.   *The Epistemological Interpretation of the Notion of a Group*

Let us first consider a general equation of the third degree

$$x^3 + ax^2 + bx + c = 0. \tag{2.9}$$

From Gauss we know that this equation has three roots $\alpha_1, \alpha_2, \alpha_3$. These three roots exist as three points in the complex plane. Using the three numbers $\alpha_1, \alpha_2, \alpha_3$ we can create a world associated with equation (2.9), namely the smallest field which contains all three roots. This field is usually represented by the symbol $Q(\alpha_1, \alpha_2, \alpha_3)$. It is the field that results when we add the three numbers $\alpha_1, \alpha_2, \alpha_3$ to the rational numbers along with everything else needed to ensure that the new system is closed under the four operations (for example $5\alpha_1 + 7\alpha_2$ and similar combinations). So far we are not interested whether this field can be constructed by algebraic means. Of course we know the answer, because we know Cardano's formulas, and so we know that the three roots can be expressed by formulas. But for a moment we will ignore this fact, and we will study the field without reference to formulas, because we would like to apply the knowledge we gain to the cases where no explicit formulas are known.

The world of algebra is a constructed world that emerged through the systematic reification of algebraic operations. Therefore we can consider the epistemic subject of algebra to be the subject which performs these operations. This subject is not identified with any particular mathematician, because a mathematician can make mistakes, while the epistemic subject is connected rather with the way that operations *should* be executed. In geometry the epistemic subject had the form of a viewpoint, and in algebra I will by analogy speak about a "viewpoint of the blind".[25] A "blind man" is also situated in his world. Nevertheless, he is usually not fixed in his world in an unambiguous way. This resembles geometry, where the viewpoint was also not given in a fixed sense, but could be moved together with a simultaneous shift of the horizon. After the world of all algebraic formulas was reified in the form of a field, the epistemic subject was not situated in this field in an unambiguous way. Its position is fixed only formally, by means of important objects like 0 and 1. In the field Q of all rational numbers the *zero* indicates where the subject "*stands*", while the *one* indicates the positive orientation and the length of his "*steps*" (if we are allowed to use these corporeal analogies). These two objects are clearly distinct, because $0 + 0 = 0$, while $1 + 1 \neq 1$, therefore the "blind man" (or the epistemic subject of the language of algebra) can distinguish them. As soon as he learns to identify these two objects, he can employ algebraic operations to construct (and so also identify) every rational number. In the world of rational numbers his position is thus unambiguously determined. Unfortunately, this field is too small, and it can only help us to solve very simple equations.

This situation changes radically when we enrich the world of the "blind man" by adding the three numbers $\alpha_1, \alpha_2, \alpha_3$, and turn from the field Q to the field $Q(\alpha_1, \alpha_2, \alpha_3)$. In this new field a fundamental problem appears: The numbers $\alpha_1, \alpha_2, \alpha_3$ satisfy the equation (2.9) and because of this the "blind man" can distinguish them from all others. Yet he cannot discriminate the three numbers $\alpha_1, \alpha_2, \alpha_3$. He knows that there are three of them, and he knows that they are different, but he cannot tell which is which. The "blind man" has no access to their numerical values, which determine the position of these three numbers on the

---

[25] As we mentioned, we consider the language of algebra a reification or objectification of motor acts. The epistemic subject of the language of algebra is thus the subject of these acts. When we call him a blind man, we have in mind not the blindness in a literary sense. We indicate only that the world is disclosed to this subject not through vision but through motor schemes. When a blind man learns how to move in a particular building, for instance in his college, he learns the different movements that bring him say from the entrance to the seminar room. The building is perhaps not the best example because it still exists also in our visually constituted world. Thus it may seem that the building is an object opened to sight and the blind men only compensate their handicap by memorizing the schemes of movement inside of this visually constituted object. In order to get an idea of the world of algebra we have to forget completely any geometrical representation of reality and imagine that there is nothing except the transformation that allows us to get from one 'place' in the structure to some other.

complex plane. Algebra allows only a finite number of steps in a calculation, while the determination of the numerical value of the roots of algebraic equations requires in general an infinite approximation procedure. From the algebraic point of view the three numbers $\alpha_1, \alpha_2, \alpha_3$ are, at least initially, indiscernible. The notion of a group was introduced in order to express this indiscernibility formally. The fact that the quantities $\alpha_1, \alpha_2, \alpha_3$ are indiscernible means that we can change their order without affecting the field $Q(\alpha_1, \alpha_2, \alpha_3)$. This field is the reification of a fragment of the language of algebra, a fragment which is invariant under any permutation of the three numbers $\alpha_1, \alpha_2, \alpha_3$. We say, that the field $Q(\alpha_1, \alpha_2, \alpha_3)$ is *symmetric* with respect to such permutations.

We can visualize this symmetry by imagining a displacement of the "blind man" in his world $Q(\alpha_1, \alpha_2, \alpha_3)$ that he cannot detect. The "blind man" cannot distinguish the $\alpha_1, \alpha_2, \alpha_3$. It may happen that he thinks he has them in front of himself in the order $\alpha_1, \alpha_2, \alpha_3$, when in reality they lie before him as $\alpha_2, \alpha_3, \alpha_1$. It is easy to see that he can err in five of the six possible ways (writing just the subscripts):

$$(1, 2, 3), (1, 3, 2), (2, 3, 1), (2, 1, 3), (3, 1, 2); (3, 2, 1). \qquad (2.10)$$

Thus the field $Q(\alpha_1, \alpha_2, \alpha_3)$, corresponding to the equation (2.9), has six symmetries. These symmetries can be combined. For instance we can, after exchanging the first two roots (creating the order (2, 1, 3)), exchange the first and the third (resulting in the order (3, 1, 2)). It is interesting that the result will be different if we make these changes in the opposite order. If we first exchange the first and third roots, yielding (3, 2, 1), and then exchange the first and second roots, the result will be (2, 3, 1).

The symmetries of a given field form a closed system under the operation of composition, which is called a *group*. A group is something like a field; it is a system of objects closed under specific operations. The only difference in our example is that while the field is formed by numbers and is closed under arithmetic operations, the objects forming the group are not quantities, but reified transformations which are closed under composition. From the epistemological point of view, a *group is a reification of a further layer of algebraic operations*, the layer of symmetries of a field. When dealing with a group we have to deal with operations on two levels. On one hand, operations are the very objects that form a group, and on the other we have the operation of their composition. Thus we could say that a group is a closed system of operations with operations.

2.2.6.2.   The Symmetry Groups of Fields Belonging to Solvable Equations

After a detour towards the notion of the group, let us return to the question of solvability of equations. We already know that the field $Q(\alpha_1, \alpha_2, \alpha_3)$, which corresponds to an equation of the third degree, has six symmetries. In order to see them more clearly, it will be useful to introduce the distinction between permutations and substitutions, which goes back to Cauchy. A *permutation*, which we will write as for instance (1, 3, 2), represents a symmetry of the field $Q(\alpha_1, \alpha_2, \alpha_3)$ in a reified form. Thus different permutations represent the different orders in which the roots $\alpha_1, \alpha_2, \alpha_3$ may be arranged. On the other hand *substitutions*, which we will write as $\left(\begin{smallmatrix} 1 & 2 & 3 \\ 1 & 3 & 2 \end{smallmatrix}\right)$, represent the same symmetries, but now in a non-reified way, as operations. This expression means that the first root stays where it is, while the other two exchange places. Thus the symbols in the upper line indicate the roots that are moving, while the corresponding symbols in the lower line indicate their destinations. The symbols (1, 3, 2)[26] and $\left(\begin{smallmatrix} 1 & 2 & 3 \\ 1 & 3 & 2 \end{smallmatrix}\right)$ represent the same symmetry, the former in reified form as the result of a transformation, the latter as the transformation itself. To each permutation there corresponds precisely one substitution, and conversely. The difference between these two notions is only an epistemological one. Nevertheless, this difference played an important role for the birth of group theory.

From the fact that one substitution corresponds to each permutation, we know that in the field $Q(\alpha_1, \alpha_2, \alpha_3)$ there are six substitutions. Thus we have a reified and a non-reified version of the group of sym-

---

[26] The objectification by means of which the notion of a *permutation* was introduced is analogous to the objectification that led to the creation of algebraic symbolism. It consists in a *replacement of an activity by its result*. A square root in the symbolic language of algebra stands for the process of root extraction and represents this process by means of its result. In a similar manner a permutation represents a process of permuting some objects by means of the final order of these objects. From an epistemological point of view the notion of a group is difficult because besides permutations it contains also substitutions and it applies substitutions to permutations. Substitutions represent a different kind of objectification than permutations. For the sake of simplicity I described them in the text as symmetries in a non-reified form. This is not absolutely precise, as each activity that is explicitly expressed in language is reified. Non-reified symmetries would be those which are shown but not expressed in the language.

metries of the field, and this made possible a clever trick. Galois asked, what would happen if we were to *apply a particular substitution to all permutations*. If, for instance, we apply the substitution $\left(\begin{smallmatrix} 1 & 2 & 3 \\ 2 & 3 & 1 \end{smallmatrix}\right)$ to the permutation (2, 1, 3), the substitution indicates that 2 will be turned into 3, 1 into 2 and 3 into 1. The result will be the permutation (3, 2, 1). Galois thus reified the permutations, combined them to form a closed system, and then investigated what would happen with this system if he applied the same substitution to all permutations. If we take the permutations

$$(1,2,3), \quad (1,3,2), \quad (2,3,1), \quad (2,1,3), \quad (3,1,2), \quad (3,2,1),$$

and apply to them the same substitution $\left(\begin{smallmatrix} 1 & 2 & 3 \\ 2 & 3 & 1 \end{smallmatrix}\right)$, the result will be

$$(2,3,1), \quad (2,1,3), \quad (3,1,2), \quad (3,2,1), \quad (1,2,3), \quad (1,3,2).$$

At first sight it may seem that the permutations have simply changed places. Nevertheless, the surprising fact is that the three permutations printed in italic characters changed their positions only among themselves, and the other three permutations again only among themselves. Thus it seems that the permutations can be divided into two blocks:

$$(1,2,3), (2,3,1), (3,1,2) \quad \text{and} \quad (1,3,2), (2,1,3), (3,2,1).$$

Galois discovered that substitutions can accomplish only one of two things: either they rearrange the permutations in the blocks, while leaving the blocks intact (as with the substitution $\left(\begin{smallmatrix} 1 & 2 & 3 \\ 2 & 3 & 1 \end{smallmatrix}\right)$), or they can exchange whole blocks. But no substitution can mix the elements between the blocks. So, for instance, no substitution can shift the permutation (1, 2, 3) into the second block, while leaving the remaining two permutations (2, 3, 1) and (3, 1, 2) in the first block. No permutation can break the borders of the blocks. The blocks either stay put or they move as wholes. Galois discovered that this respecting of the boundaries of the blocks of permutations by the substitutions is closely related with the fact that the equation of the third degree is solvable. The reason for this is that the three roots $\alpha_1, \alpha_2, \alpha_3$ can be expressed by the use of the four arithmetic operations plus root extraction. Any field created in this way has symmetries which can be divided into blocks, and these blocks into further blocks so on, so that at the end we will come to a block, the number of elements of which is a prime number (in our case we got a block with three elements). In other words the symmetry group of fields, which corresponds to solvable equations, can be factorized into cyclic factors.

2.2.6.3. *Insolubility of the Equation of the Fifth Degree*

By reifying the symmetries of the particular fields, Galois reached a level of abstraction that allowed him to understand why the quintic equation is in general insoluble. Gauss had already showed that every equation of the degree n has n roots. Therefore each quintic equation has five roots $- \alpha_1, \alpha_2, \alpha_3, \alpha_4, \alpha_5 -$, and the problem is that these roots cannot be expressed by algebraic means. The integrative form of language makes it possible to understand why this is so. It enables us to replace the question, whether the roots $\alpha_1, \alpha_2, \alpha_3, \alpha_4, \alpha_5$ can be expressed by algebraic means, by the question whether the group of symmetries of the field $Q(\alpha_1, \alpha_2, \alpha_3, \alpha_4, \alpha_5)$ can be split into smaller and smaller blocks. This last question is not so difficult to answer. It is sufficient to take all the permutations of five elements (of which there are 120), and see what happens to them under different substitutions. The case of five elements is more complicated than the case of three elements, which we discussed above, but these difficulties are not fundamental. Galois discovered that the only possible division into blocks is a division into two blocks containing 60 elements each. But if we restrict ourselves to one of these two blocks, we have a group with 60 elements, which is called the *alternating group of five elements*. This group cannot be further divided into blocks, because the permutations mix the elements between any blocks. The discovery of this fact was one of the most surprising moments in the history of algebra.

The symmetry group of every field that can be constructed by algebraic means can be factored into a system of nested blocks. Galois discovered that the symmetry group of the field $Q(\alpha_1, \alpha_2, \alpha_3, \alpha_4, \alpha_5)$ associated with an equation such as $x^5 - 6x + 3 = 0$ cannot be factored in such a way. This means that no field constructed by algebraic means can ever contain the roots of this equation. Therefore there cannot be any general formula for the solution of quintic equations analogous to Cardano's formulas for cubic equations. This shows that the solvability of equations in terms of radicals is an exceptional phenomenon. Only equations with special symmetry groups turn out to be solvable. Cubic equations, for instance, are solvable because the associated fields only have six symmetries, which can be divided into two blocks. Beginning with the equation of the fifth degree, however, no such division is possible, and therefore there is no formula capable of solving this equation. The universe of algebraic formulas is too simple. It does not allow us to construct fields with symmetries complex enough to encompass the field $Q(\alpha_1, \alpha_2, \alpha_3, \alpha_4, \alpha_5)$ associated with the fifth-degree equation. Therefore algebra has to shift its focus from formulas to *algebraic structures*. Algebraic structures, as for instance groups, decide what can be expressed by formulas and what cannot. The discovery of the alternating group of five elements marks the start of modern structural algebra.

2.2.7.   The Constitutive Form of the Language of Algebra

The integrative form of language introduces a neutral basis in which different structures are embedded. These structures are then compared by means of group theory. Cayley and Klein used this approach in geometry where the neutral basis was the projective plane. Galois in his study of algebraic equations supposed that the roots $\alpha_1, \ldots, \alpha_n$ of a particular equation were given as complex numbers, and thus the field $Q(\alpha_1, \ldots, \alpha_n)$, corresponding to the equation, was embedded in the field of all complex numbers. In this way the field of complex numbers played a role in algebra similar to that played by the projective plane in geometry. In both cases the next step consisted in the elimination of the neutral basis, in *making the structure independent of its basis*. In geometry Riemann removed the condition of the construability of a particular object in three-dimensional space as the criterion of the object's existence, and created combinatorial topology in which the objects are studied independently of whether they can or cannot be constructed in space. Thus Riemann eliminated space, which until then had been the basis of geometry (be it the projective space, or any other), instead using language as the basis upon which the existence of objects was constituted.

The analogous move in algebra consisted in making the notion of group independent of an underlying field, the symmetries of which the group represents. It was the German mathematician Heinrich Weber who around 1893 discovered how to do this. When constructing the extension of the field Q by adding to it an element α, Weber decided not to "borrow" this element from the complex plane, as Galois had done. Rather he sought to add the element α to the field Q by purely algebraic means. At first sight this may look strange, because if α is the root of the equation $x^5 - 6x + 3 = 0$, we know that it cannot be represented by any algebraic formula. Thus one might well ask what it means to add an element α to the field Q by purely algebraic means, when this element cannot be represented by the means of algebra. And here Weber used the analogy with the introduction of the imaginary numbers. There too we had equations that could not be solved by any number. And what did the algebraists do? They extended the language and introduced the square roots of negative numbers as a kind of new symbols. So why not to do the same thing here? Weber's construction is a bit lengthy and so we will divide it into its basic steps.

2.2.7.1. Construction of the Ring $Q[x]$

Let us take the field Q, which we want to extend, and add to it an uninterpreted symbol, for instance x (in the case of complex numbers it was i, so now we take x). In this way we obtain the system $Q[x]$, which is called the ring of all polynomials over the field Q. It is another kind of closed world where we can add, subtract, and multiply any two of its members. A ring differs from a field in that in a ring it is not possible to divide. When we divide a polynomial by another polynomial, the result will not always be a polynomial, thus the ring is not closed with respect to division. So the ring $Q[x]$ consists of all polynomials of all possible degrees, the coefficients of which belong to Q:

$$Q[x] = \{a_n x^n + a_{n-1} x^{n-1} + \ldots + a_1 x + a_0; a_i \in Q, n \in N\}.$$

The construction of the ring $Q[x]$ was Weber's first step. The ring $Q[x]$ differs from a field of the kind $Q(\alpha)$ in two respects. The ring is not, as stated, closed under division, and the element x is not bound by any conditions, while the number α was a root of the original algebraic equation. The next step is to modify the ring $Q[x]$ so that it will resemble the field $Q(\alpha)$. We have to bind the element x by further conditions in such a way that it will mimic the number α. In this way we obtain a field isomorphic to $Q(\alpha)$, but without any recourse to the complex plane. The tool for this step is the notion of an ideal.

2.2.7.2. Construction of the Ideal $(g(x))$

Let us recall what we are doing when we add an element to the field Q. Let us take for instance the number $\sqrt[3]{2}$. We have to construct all the possible expressions containing this number. First of all, all the linear ones, such as $(5. \sqrt[3]{2} + 7)$, then all the quadratic ones, such as $(3 \cdot (\sqrt[3]{2})^2 - 17 \cdot \sqrt[3]{2} + 6)$. Actually, we do not need to go further because the third power of $\sqrt[3]{2}$ is simply 2. For the algebraic calculations we do not need the exact numerical value of the number $\sqrt[3]{2}$, it is sufficient to know that its second power is independent from it, while its third power is 2. All this we can gather from the expression $\sqrt[3]{2}$ itself, but the same information can also be obtained from the equation $g(x) = x^3 - 2$, of which $\sqrt[3]{2}$ is a root. Thus for algebraic purposes the field $Q(\sqrt[3]{2})$ can be viewed as a $Q(\alpha)$ for some α such that $\alpha^3 - 2 = 0$. In the case of a polynomial of the fifth degree, for instance $g(x) = x^5 - 6x + 3$, we have no expression of its roots (analogous to $\sqrt[3]{2}$), but Weber seeks nevertheless to construct a field $Q(\alpha)$ for some α. The only difference will be that instead of requiring $\alpha^3 - 2 = 0$ he requires $\alpha^5 - 6\alpha + 3 = 0$. In other words he requires that α be a root of the polynomial $g(x)$. Of course, we have such an α at our disposal, namely a certain complex number. But Weber did not want to make use of non-algebraic means. He wanted to construct α by purely algebraic means. The basic idea for such a construction comes from Dedekind.

The simplest way to explain what an ideal is is to turn to the integers (which form a ring but not a field, just like the polynomials, because the division of two integers may not be an integer). Dedekind introduced the notion of an ideal as a system of numbers closed under addition and multiplication by an integer. An example of an ideal in the ring of integers is for instance all multiples of the number 6, because the sum of any two such numbers as well as an integer multiple of any such number is again a multiple of the same number 6. We write

$$(6) = \{6a; a \in Z\} = \{0, 6, -6, 12, -12, 18, -18, \ldots\}.$$

Since 3 divides 6, all multiples of 6 are also multiples of 3. This fact can be succinctly expressed in the language of ideals by saying that the ideal (6) is contained in the ideal (3). Of course, this is trivial. The nontrivial aspect of the theory of ideals is that in some cases there exist ideals that are not simply collections of all the multiples of a single element. This means that *the language of ideals is richer than the language of numbers* (not in the case of ideals of integers, but in the more general cases). Thus we can associate an ideal with each number, but there may exist ideals that cannot be represented in this way, ideals to which no element in the realm of numbers corresponds. This was the reason why Weber turned to the theory of ideals. Instead of the ring of integers he worked in the ring of polynomials, but the basic aim was the same, namely to get access to a richer realm of objects.

Similar to associating with the number 6 the ideal (6) consisting of all (*integer*) multiples of 6, Weber associated with the polynomial $g(x)$

an ideal consisting of all (*polynomial*) multiples of $g(x)$:

$$(g(x)) = \{g(x) \cdot f(x); f(x) \in Q[x]\}$$
$$= \{g(x) \cdot 2x, g(x) \cdot x^2, g(x) \cdot (x^3 + 7x + 3), \ldots\}.$$

By means of this ideal he was able to construct abstract objects representing the solutions of equations that cannot be solved by algebraic means. He achieved this by means of factorization.

2.2.7.3.   *Factorization of the Ring $Q[x]$ by the Ideal $(g(x))$*

Weber's aim was to extend the field Q by an object, which would mimic the root of the polynomial $g(x)$. Let us represent this extension as L (we don't want to use the symbol $Q(\alpha)$, because the complex number α used by Galois does not belong to the language of algebra). In order to obtain the field L, Weber used a construction analogous to the construction of residue classes in number theory. In number theory a residue class consists of all integers which have the same remainder after division by the given number. Thus for instance the class $\overline{3}$, consisting of all integers, which give the remainder 3 after division by 6 can be written as

$$\overline{3} = \{3 + 6a; a \in Z\} = \{3, -3, 9, -9, 15, -15, \ldots\}.$$

We can do the same thing with polynomials. In the case of a given integer we only have a finite number of residue classes, while in the polynomial case the number of residue classes is <u>infinite</u>. One such class in the polynomial case is, for instance, the class $3x + 2$ of all polynomials that give after division by $g(x)$ the same remainder $(3x + 2)$:

$$\overline{3x + 2} = \{(3x + 2) + g(x) \cdot f(x); \quad f(x) \in Q[x]\}$$
$$= \{(3x + 2) + g(x) \cdot 2x, (3x + 2)$$
$$+ g(x) \cdot x^2, (3x + 2) + g(x) \cdot (x^3 + 7x + 3), \ldots\}.$$

Weber showed that all the residue classes created as classes of polynomials having the same remainder when divided by an irreducible polynomial $g(x)$, *form a field*. Thus we can add, subtract, multiply, and divide such classes. The greatest surprise was that this field of residue classes is isomorphic to the field $Q(\alpha)$, which Galois constructed with the help of the complex number α. Thus Weber succeeded in constructing the field L without leaving the language of algebra. He constructed the field L from the ring $Q[x]$ as the system of all residue classes created by dividing by the polynomial $g(x)$.

The element which in Weber's field L corresponds to the element α of the extension $Q(\alpha)$, i.e., the element representing the root of the equation $g(x) = 0$, is the residue class

$$\bar{x} = \{x + g(x) \cdot f(x); f(x) \in \mathbf{Q}[x]\}$$
$$= \{x + g(x) \cdot 2x, x + g(x) \cdot x^2, x + g(x) \cdot (x^3 + 7x + 3), \ldots\}.$$

It is the residue class that contains all polynomials which when divided by $g(x)$ give the remainder x. To see that this abstract object $\bar{x}$ is a root of the equation $g(x) = 0$, it suffices to substitute any member of the class $\bar{x}$ for x in this equation. In so doing, we have to remember that we are working in the system of residue classes, so that $\overline{g(x)}$ is the same class as $\bar{0}$, as is indicated by the equation $g(x) = 0$.

Weber's construction indicates the strength of the theory of ideals. This theory makes it possible to construct objects which cannot be represented in the language of classical algebra. Weber constructed the field L which has all the properties of the field $Q(\alpha)$, and thus contains the root of the insoluble polynomial. This root – the class $\bar{x}$ – is given as an *element of an abstract structure*, and not by means of a formula. Galois' result about the insolubility of the quintic equations remains valid, but Weber's construction sheds new light on this result. It shows that the problem lies in the restriction of the means employed. If we restrict ourselves to the four arithmetic operations and root extraction, the quintic equation is insoluble. But if we allow more abstract means, every equation will be solvable in this new, abstract sense. Weber introduced into algebra the constitutive form of language.

Analyzing Riemann's work in geometry, we have seen that the constitutive form of language liberated the objects from their confinement in geometrical space and that the role of constituting the existence of geometrical objects was transferred from space to language. Due to the new form of language geometry obtained access to objects that were formerly incomprehensible, as for instance the *Klein bottle*. Weber accomplished something similar in algebra. He *liberated the objects of algebra from their dependence on formulas*. Before Weber, algebra based the existence of its objects on the possibility of their formal representation. Thus in a sense algebraic objects were contained in the "space" of formulas. This was the reason why such objects as the roots of the polynomial $x^5 - 6x + 3 = 0$ had been unimaginable in algebra. Weber changed this situation when he showed that every algebraic polynomial can be solved by algebraic means. Weber's factorization is similar in many respects to Riemann's construction of surfaces. Just as Riemann identified particular parts of the border of a surface, Weber identified all polynomials that yield the same remainder when divided by a given ideal. Thus factorization can be seen as a *constitutive act*, which makes the construction of new objects possible. In Weber's approach every polynomial is solvable. The only difference is that the roots of the polynomials are more abstract objects than numbers.

2.2.8. The Conceptual Form of the Language of Algebra

In his paper (Weber 1893) Weber introduced a new definition of the notion of a group. It is the first definition which included groups with an infinite as well as a finite number of elements. Before Weber mathematicians studied only finite groups which can be defined using the law of cancellation (i.e., by the requirement that $a \cdot x = b \cdot x$ entails $a = b$). For the infinite case this definition does not work, and Weber replaced it by the requirement of the existence of inverses (i.e., that for each element x of a group there is an element x^{-1} such that $x \cdot x^{-1} = 1$). Weber's motivation for extending the concept of a group to include the infinite case was that he wanted to use the notion of group as a basis for the definition of a field: *"From a group a field emerges, when there are two ways of composition in it, one of which is called addition, the other multiplication."* So Weber was one of the first mathematicians to view a field as we do today, namely as a group, into which a supplementary operation is introduced.

This move fundamentally changes the epistemological relation between the notions of a group and of a field. In the integrative form of the language of algebra a field was a world and a group was the structure of its symmetries. That meant that the notion of the field was primary while that of a group was derived. Weber obliterates this epistemological difference. From the point of view of the conceptual form of language all notions are on the same level. Notions are defined by axioms, and the only thing that matters is the logical interdependence among them. As the definition of a field presupposes the notion of the group, because each field contains an additive group, from the viewpoint of the conceptual form of language the notion of a group is prior to the notion of a field.

The gain brought by Weber's introduction of the conceptual form of language into algebra is obvious. The explicit formulation of all the conditions in the definitions of the fundamental notions opens the possibility of weakening some of these conditions, and in this way to create different generalizations. From the notion of a group, for instance, we can create the notions of a *groupoid*[27] or of a *semi-group*, notions whose introduction would either be inconceivable or at least much more difficult to motivate at the previous stages. If we do not introduce our concepts "from below" through stepwise generalizations of particular examples, but rather accept their introduction "from above" by changing the conditions in the definitions of the notion of group, we get much easier access to them. In this respect the conceptual form of language opened up a totally new realm of algebraic notions. Thus another role of language, the role of defining concepts, becomes reified. The previous stages in the development of algebra used only "natural notions", notions which appeared in a natural way in mathematical research. In contrast to this the conceptual form of language studies concepts independently of any natural context. It studies them systematically by varying the definitions of already existing notions. This development is again not specific to algebra, and its parallel can be found in topology, where in a similar manner, after the notion of a topological space was introduced, a whole array of different kinds of topological spaces emerged. Thus the conceptual form of languages offered a freedom in concept formation, which was previously inconceivable.

2.2.9. An Overview of Relativizations in the Development of Algebra

Weber's factorization makes it possible to solve all algebraic equations. So the prophecy of Viète was fulfilled – no problem remained unsolved. Let us end the reconstruction of the development of algebra here. I took up the problem of the solution of algebraic equations as a kind of a thread to lead us through the labyrinth of the history of algebra. Instead of going further I would like to summarize our results. We have discerned seven forms of language of algebra, which differ in the way they conceive of a solution of algebraic equations. To solve an equation means:

---

[27] An interesting discussion of the meaning and importance of the notion of a groupoid can be found in (Corfield 2003, p. 208–230).

1. For the perspectivist form – to find a *regula*, i.e., a rule written in ordinary language, which makes it possible to *calculate* the "thing", that is, the root of the equation.

2. For the projective form – to find a *formula*, i.e., an expression of the symbolic language, which makes it possible to *express* the root of the equation in terms of its coefficients, the four arithmetical operations, and root extraction. The symbols in the formula correspond to steps of the calculation, and so a formula is a representation of the regula.

3. For the coordinative form – to find a *factorization* of the polynomial form, i.e., to *represent* the polynomial form as a product of linear factors. Each factor represents one root of the equation, and so the number of the factors is equal to the degree of the equation.

4. For the compositive form – to find a *resolvent*, i.e., to *reduce* the given problem, by means of a suitable substitution, to an auxiliary problem of a lesser degree. A solution to the auxiliary equation can be transformed into a solution of the original problem. But here, besides the n roots of the nth degree equation we also obtain all the associated quantities, in general, $n!$ numbers.

5. For the interpretative form – to find a *splitting field*, i.e., to *construct* the field $Q(\alpha_1, \ldots, \alpha_n)$ that contains all the roots of the equation. This field automatically contains all the associated quantities of the equation, and thus also all the roots of its resolvent.

6. For the integrative form – to find a *factorization* of the Galois group of the splitting field $Q(\alpha_1, \ldots, \alpha_n)$, i.e., to *decompose* the group of symmetries of this field into blocks. Steps in the factorization of the group correspond to particular extensions of the field. Hence from knowledge concerning the factorization of the group we can draw conclusions about the field extensions.

7. For the constitutive form – to construct a *factorization* of the ring of polynomials $Q[x]$ by the ideal $(g(x))$, i.e., to *find* the residual classes of the ring of polynomials after factorization by the ideal that corresponds to the equation we want to solve. One of these classes is the root of the equation, and so we have an abstract method of solution of algebraic equations.

Besides these differences on the *intentional* level, the particular forms of language differ also with respect to their *ontology* and *semantics*.

Thus their discrimination can be seen as a first step towards a better appreciation and understanding of the richness of philosophical issues that we encounter in algebra. The process of gradual reification of operations described in this chapter can be viewed as a contribution to the discussions about the ontology of mathematics. It seems that each form of language of algebra has ontology of a specific kind. To develop a philosophical account of mathematical ontology would therefore require an account of the common aspects as well as of the differences among the ontologies of the particular forms of language.

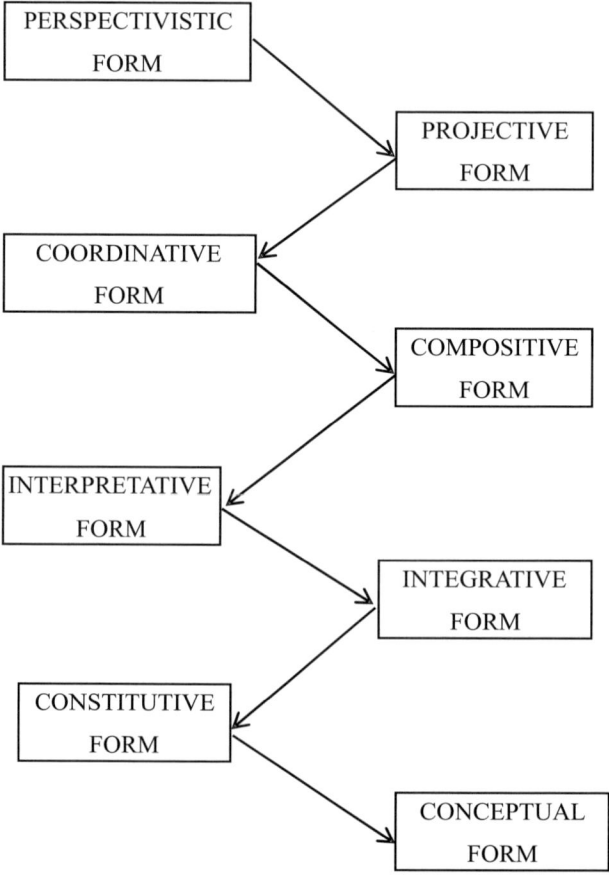

2.3. Philosophical Reflections on Relativizations

The changes of language described in the second part of the present book are fundamentally different from re-codings. While *re-codings* introduce *new ways of generating* pictures (in synthetic, analytic, or iterative geometry) or formulas (in arithmetic, algebra, differential and integral calculus, or predicate calculus), relativizations take place within a particular representation. Relativizations do not change the way of generating linguistic expressions. *Relativizations* change rather the *ontological commitments* of language, and the way that the *reference* of the linguistic expressions is understood.

As an illustration let us take the changes related to the introduction of the projective form of language. The first of these changes consisted in giving the background the same *ontological* status as the objects had. Thanks to this the background becomes incorporated into the language of the theory. In geometry the background has the form of the space in which the different objects take up their position; in algebra it is the system of quantities which represent the possible values for the variables. The second important change is the explicit representation of the *viewpoint* in the theory and the following relativization of the objects with respect to this viewpoint. In this way the *homogeneity* of the language is broken. In the language, special expressions appear (the center of projection and the horizon in geometry, the zero and the one in algebra) that represent the epistemic subject. Further the projective form of language introduces a *representation of representation* and so a duplication of expressions of the language. In this way two systems of expressions appear: the transformation and the transformed object in geometry; the operation and the result of the operation in algebra.

We can represent the world "from nowhere" as Euclid did, or we can represent it from a particular point of view as Desargues did. It is also possible to model it, as Beltrami did, and it is possible to characterize it by the constitutive acts that are necessary in its construction, as Poincaré did. Just as in the case of re-codings, so also in the case of relativizations we will turn after the reconstruction of history to a discussion of the philosophical and educational implications of this pattern of change in the development of mathematics.

2.3.1. Comparison of the Development of Algebra with the Development of Geometry

So far we were moving on the safe ground of historical reconstructions. Now I would like to discuss some more general ideas of an epistemological nature. Their purpose is to offer a unified interpretation of the epistemological changes that occurred in the development of algebra and of geometry. The first step must be therefore a comparison of the results that we obtained in reconstruction of the development of geometry and algebra.

2.3.1.1.   *Omission of Some Forms of Language in the Development of Geometry*

Perhaps the most important contribution of the reconstruction of the development of algebra was the discovery of the *coordinative and the compositive forms of language*. While these forms are clearly present in the development of algebra, in the reconstruction of the development of geometry I could not find them. Thus one kind of differences between the development of geometry and algebra may consist in the fact that some of the eight forms of language that I discussed above can play no role in the development of one of these disciplines, while it has an important role in the development of the other. This sheds some light on our method of reconstruction of the semantic development of mathematical theories. The list of eight forms of language represents a succession of different possibilities of how the correspondence between the intended domain of the theory and its expressions can be arranged. These forms are ordered successively according to their growing complexity. If in the use of the theory some difficulties appear, one of the possibilities for solving them is to turn to a language with a more complex form, a form that is able to represent more complex situations. Therefore the most natural thing to do is to take one step in the succession of forms presented in our list. Nevertheless, it may happen that the next form of language, that is the language with the smallest possible complication of semantic structure, is not appropriate for the solution of the particular problems which the theory encountered. So it may be necessary to make a more radical change, which will be manifested in omitting one or two stages in our list and turning to some more elaborate form of language. It seems that this happened in the development of non-Euclidean geometry, when the coordinative form was of no use for the problems encountered by the founders of these new geometries, and so they turned directly to the interpretative form. Therefore our list offers a maximal system of stages, a complete (at least I hope so) system of possible arrangements of the semantic correspondence of expressions of the language with its intended universe, and from this list the actual development can select the appropriate ones. It is possible that in the reconstruction of the development of other areas of geometry (for instance of analytic geometry) the coordinative form would play an important role[28] and some other form of language would not be used.

2.3.1.2. Alternation of the Implicit and Explicit Variants of the Form of Language in Geometry

An interesting aspect of reconstruction of the development of geometry, which has no counterpart in the development of algebra, was a regular alternation of *implicit* and *explicit* forms. This was possible, because in geometry each pictorial form exists in these two clearly distinct versions. The situation in algebra is much more complicated. If we look at the development of algebraic symbolism, we see that it is a rather slow and gradual process, stretching from Regiomontanus to Descartes, over more than two centuries. The dynamics here consists in a slow process of reification of expressions of the language, instead of a change from an implicit version of the form to its explicit version (as was the case in geometry). The world of geometry is opened as a whole to our sight, and therefore any change it undergoes must happen at once, as a *Gestalt switch*. In algebra, by contrast, the world that is given to us is only fragmentary, we know only some of its "places" where we have "fumbled around" (in our calculations), we know only a few "tricks" which we have found (such as the substitution $x = \sqrt[3]{u} - \sqrt[3]{v}$). Thus in algebra the emergence of a new pictorial form happens slowly and gradually, and does not resemble a Gestalt switch.

The contrast between the world opened to our sight and the world constituted in the process of reification of linguistic expressions en-

---

[28] The barycentric calculus of Möbius may be based on the coordinative form of language and so it is possible that the reconstruction of the development of analytic geometry will also require the use of that form. Similarly it is possible that a more detailed analysis of the work of Saccheri will disclose the use of the compositive form of language. Thus it may happen that also in the development of geometry all forms that occurred in the history of algebra have been used. Nevertheless, what I would like to stress is that from the theoretical point of view it is not necessary that the stages in different areas should be in one-to-one correspondence.

ables us to explain another peculiarity of algebraic texts. Let us take Cardano's *Ars Magna*. When we take up this book, we discover that it contains fragments belonging to different forms of language, fragments having very different semantic structures. It contains the rules for the solution of cubic equations. These rules are formulated in ordinary language, which is characteristic of the *perspectivist form*. As we have shown, these rules were derived with the help of a substitution, which is a typical feature of the *projective form*. Further, the book contains the *casus irreducibilis*, which is the germ of a new, *coordinative form*, the form in which the complex numbers will be incorporated into algebra. Thus it seems that an algebraic text may contain different fragments, belonging to three different forms of language. In geometry such a co-existence of fragments belonging to different forms is inconceivable. It is impossible to have a picture which is partly Euclidean and partly non-Euclidean. In geometry the world is disclosed as a whole, and thus all its parts must fit together.

This difference is very important from the epistemological point of view. It shows that the explicit and implicit versions of a form of language, which we discussed in our analysis of geometry, are only two ways of expressing the same thing. It is a peculiar feature of geometry that each form of language exists in these two versions – explicit and implicit – and that their alternation happens in a regular pattern. This led me in (Kvasz 1998), erroneously as it now appears, to assume that the dynamic of alternations of the implicit and explicit forms of language is the epistemological dynamics behind relativizations. In algebra each form of language exists in several versions which form a gradual transition from the fragmentary, through the implicit to the explicit version. Even after the transition to the explicit form of language, some fragments of the previous form still survive. I therefore came to the conclusion that the alternation between implicit and explicit versions of the form of language, which was such a remarkable feature of the development of geometry, does not constitute the dynamics of relativizations. The fragmentary, implicit and explicit versions of a form of a language are only its different formulations, and so their changes constitute the dynamics of re-formulations. It therefore seems reasonable to define the notion of the form of language in a broader sense, and to include its fragmentary, implicit, and explicit versions in a particular form of language.

2.3.2. Form of Language and the Development of Mathematical Theories

In the course of our historical reconstruction of the development of geometry and of algebra, some aspects of the language of mathematics were systematically recurring. In the case of geometry these were *the point of view* (from which a particular figure has to be viewed in order to understand it correctly), *the horizon* (corresponding to the boundary of the world represented by the picture), *the background* (the plane or space into which the geometrical objects are situated), and *ideal objects* (such as the infinitely remote points in which the parallels meet). It is interesting to realize that the fundamental changes that occurred in the development of geometry concerned first of all these aspects. The reconstruction of the development of algebra has shown that these four aspects have to be complemented by a further two: *the individua* and *the fundamental categories*.

2.3.2.1. Specification of the Notion of the Form of Language

I suggest introducing the notion of the form of language as a structure consisting of six related aspects:

the epistemic subject of the language,   *the horizon of the language,*
the individua of the language,   *the fundamental categories of the language,*
the ideal objects of the language,   *the background of the language.*

I believe that these aspects are formal, i.e., they have no factual meaning. Let me explain what I mean by this with the example of the *horizon*. If we take a perspectivist painting of a landscape, we can clearly recognize a line, which is called the horizon. Nevertheless, if we went out in the countryside, represented by the painting, to the place of the alleged horizon, we would find nothing particular there. And the painter, when painting his landscape, did not paint the horizon by a stroke of his brush. He painted only houses, trees, hills, and at the end the horizon was there. This is the meaning of Wittgenstein's words: "*A picture cannot depict its pictorial form: it displays it.*" The painting does not depict the horizon; it displays it. The horizon is an aspect of the form of language and that means that it cannot be empirically determined. Despite the fact that in the picture the horizon can be clearly determined, in the world represented by the picture there is no object corresponding to it. It is interesting that the languages of mathematical theories are full of such non-denotative expressions. Take for instance zero in different algebraic structures, the negative or the complex numbers. There is no empirical procedure that would allow us to determine the denotation of these terms.

When we include into the form of language the individua and the fundamental categories, the overall structure of the notion of form of language becomes clearer. Its purpose is to connect the subject (the user of the language) with his world (the universe of the language). This connection happens on three levels. The first function of the form of language is to *incorporate the epistemic subject into the world*. This is achieved by means of the point of view and of the horizon. *The point of view* (in projective geometry it was the centre of projection, in algebra the number zero) indicates the position of the subject, from the viewpoint of which the theory is formulated. The point of view thus incorporates the speaker into the world; it constitutes the *identity of the subject* in the universe of the language. *The horizon* (in projective geometry the vanishing line, in algebra the unit) coordinates the world and the epistemic subject. When it fixes the basic directions (the vanishing line determines the horizontal plane and in this way it determines the directions upwards and downwards; the unit determines the positive direction of the number line, thus discriminating the increasing from the decreasing), it constitutes the *situatedness of the subject* in the universe of the language.

The second function of the form of language is to *structure the world from the point of view of the epistemic subject*. This is achieved with the help of the individua and of the fundamental categories of the language. To determine *the individua* means to identify objects in the world that are in a sense analogous to the subject; objects that the epistemic subject can refer to in a fixed way. Not accidentally the term "body", by which individua are designated in classical mechanics, has its roots in the English term for corpse. A body in mechanics is something analogous to our human body, something we can refer to in a corporeal way. *Individuality* is a fundamental attribute of the subject. The subject encounters himself as an individuum, and projecting his individuality to certain objects, he constitutes them as individua of the language. This is why the determination of the individua cannot be an empirical question, which could be decided independently from the form of language. Thus for instance in projective geometry, incorporating the infinitely remote points into the language, the two parts of a hyperbola are considered as forming a single curve, i.e., an individuum. Earlier it was natural to consider them as two different objects, i.e., as two individua. On a more abstract level even the three conic sections, the ellipse, the parabola, and the hyperbola, can be seen as three different positions of the same object, i.e., as a unique individuum. The next step to make after language has divided the homogenous continuum of being into discrete individua is to introduce different *similarity relations* into the world. Congruence, similarity, or affinity are introduced into the set of all geometric figures; similarly different congruence relations are introduced in algebra. In this way language introduces certain structures into the world, and so it makes the world more familiar for the subject.

The third function of the form of language – besides incorporation of the subject and the structuring of the world – is to introduce a homogenous background to the world. This helps the subject to *find his orientation in the world.* Typical examples of this are the different kinds of space in geometry (projective space, affine space, metric space, topological space), the different number systems in algebra (the system of all natural numbers, the system of all real numbers, different fields, rings). They do not refer to anything real, so we could possibly avoid using them. But this would only make our language more cumbersome. It is useful to add these formal objects to our language, which can help us to find our *orientation* in the world. On this homogenous background we build different schemes. Nevertheless, it might happen that the world is not in accordance with our schemes. Then it is useful to add some ideal elements to the language, just as the infinitely remote points are added to space in geometry, or the complex numbers are added to the number system in algebra. These again are expressions of the language, having no real denotation and therefore we include them into the form of language. But if we add them to the language, the world becomes more *transparent.* Schemes, which had only a restricted validity, become universal. After having added the infinitely remote points to the plane, any two straight lines will have an intersection. Thus in the proofs it is not necessary to distinguish the different cases depending on the intersection of lines. Similarly in algebra after having added the imaginary numbers to the number realm every number will have a square root, and so it is not necessary to distinguish the different cases, depending on whether an expression is positive or not. The schemes work more smoothly, the world is more transparent.

2.3.2.2.   The Development of Mathematical Theories Seen as a Series of Changes of the Form of Language

We have seen that the purpose of the respective aspects of the form of language is to situate the epistemic subject into the universe of the language, to structure this universe in a way comprehensible for the subject, and to provide means for a better orientation in this universe. The aspects of the form of language, which constitute the identity, situatedness, individuality, similarity, orientation, and transparency, are not factual. The subject does not belong to the world (*Tractatus* 5.632: *The subject does not belong to the world: rather, it is a limit of the world*), the horizon does not denote anything factual, there are no fix individua, space does not really exist, not to speak about the ideal elements. But this is exactly the reason why the different aspects of the form of language are an ideal tool for the description of relativizations, i.e., of the kind of changes that are analyzed in this chapter. Those elements of language that refer, in a direct or indirect way, are bound by the relation of reference and therefore they cannot change with sufficient flexibility. On the other hand the aspects of the form of language, precisely because they do not refer to anything real in the world, are free. They are bound only by relations among each other. Therefore in cases when the development of the discipline requires a change, the aspects of the form of language offer sufficient space for innovation.

To avoid misunderstandings, I would like to stress that my approach to the development of mathematics, despite its use of historical reconstructions, has nothing to do with historicism. I do not consider the regularities in the development of geometry and of algebra to be some historical laws. My approach is rather a version of structural analysis and I use historical material only because in history we can find a great variety of different languages used in the same area of mathematics. The regularities in the development of mathematics can be explained using three principles:

1. There is only a limited number of ways in which the expressions of a mathematical language can be brought into correspondence with the universe of discourse in a consistent manner. These ways can be characterised by the notion of the *form of language*. Until now we have found *eight* of these forms.

2. When in the development of a theory some deep problems appear, one possibility for dealing with them is to *change the form of*

language. The new form often brings means that allow us to solve problems insoluble on the basis of the previous form.

3. Scientists are conservative and thus they try to make the smallest possible change in language.

These principles introduce regularity into the historical succession of the forms of language. The succession of forms that we have described (*perspectivist form, projective form, coordinative form, compositive form, interpretative form, integrative form, constitutive form, conceptual form*) is ordered according to the growing complexity. The source of regularity in the development of mathematical theories is thus not some historical necessity but the rationality of the mathematicians who reach after a more complex form only after they have exhausted all the possibilities of the simpler forms. In this way the common regularity that we discovered in the development of geometry and of algebra can be explained. The developments of geometry and of algebra show a common pattern (with the exceptions that were discussed in Section 2.3.1) because mathematicians tried to solve the difficult problems that occurred in these disciplines with the smallest possible changes of the structure of the language of mathematics. Thus even though we discovered this common pattern as a temporal regularity, i.e., as a pattern of alteration of the forms of language in the historical development of geometry and of algebra, the origin of this pattern is purely structural. It is the consequence of the fact that there is a limited number of different forms of language which can be linearly ordered according to the growing complexity of their epistemological structure. It was this linear order of growing complexity that gave rise to the common temporal pattern in which these forms of language appeared in the historical development of geometry and of algebra. We see that the notion of the form of language is a useful tool for the reconstruction of the history of different mathematical disciplines.

We can thus formulate a research program to reconstruct in a similar way the development of other mathematical disciplines. In the field of geometry (or iconic languages, as they were called in chapter 1) we can try to find a reconstruction also for analytic, algebraic, differential, and fractal geometries. Similarly in the area of symbolic languages we can try to reconstruct besides algebra also arithmetic, the differential and integral calculus, and predicate calculus. As a further application of our approach we can attempt a similar reconstruction of the development of classical mechanics, thermodynamics, or quantum theory.

2.3.3.   The Notion of the Form of Language and Philosophy of Mathematics

Another area for application of the notion of form of language, besides in reconstruction of the development of different mathematical theories, is of course the philosophy of mathematics. Already in point *c* of Chapter 2.1.4.3 I tried to connect Poincaré's philosophy of geometry with the integrative form of language. It seems plausible that Poincaré's views on geometry expressed in his *La Science et l'Hypothése* (Poincaré 1902) offer a deep philosophical analysis of the situation that appeared in geometry after the introduction of the integrative form of language. In a similar way it would perhaps be possible to connect Kant's philosophy of geometry with the compositive form, and that of Descartes with the coordinative form. If we would follow this idea in a systematic way, i.e., if we understand a particular philosophical interpretation of mathematics not as a universal theory, but try to situate it on the background of a particular form of language, it could help us to understand better the different debates in the philosophy of mathematics. We could perhaps see more clearly both: the area of the positive contribution of a particular school of philosophy (which often consists in a better understanding of achievements in mathematics made on the basis of a new form of language) as well as the school's limitations (which are often caused by insensitivity to the problems and techniques brought about by some further form of language).

2.3.3.1.   Philosophies of Mathematics and Their Relation to the Form of Language

The relation of particular philosophies of mathematics to corresponding forms of language could help avoiding strict condemnations of some positions which (similarly to Kant) are perfectly adequate and legitimate when restricted to a particular form of language; and at the same time this could guard us against the temptation to defend the indefensible (as it sometimes happens in connection with Kant) and to see clearly the point when it is necessary to give up a philosophical view and replace it by some other. If we interpret Kant's philosophy of geometry as a position based on the compositive form of language and Poincaré's philosophy of mathematics as a position based on the integrative form of language, a new approach to philosophy of mathematics arises. It seems that the *scope of validity of a particular position in the philosophy of mathematics* can be most precisely characterized using the notion of the form of language. Many controversies in the philosophy of mathematics have their roots in a *change of the form of language* in mathematics. By means of the new form mathematics arrives at new results, which contradict the considered philosophical position. Nevertheless, because the form of language cannot be expressed but it only shows itself, the mathematicians and the philosophers are usually unaware of the change of the form of language, and so they consider that the new mathematical discoveries should lead to discarding the original philosophical position. I am convinced that this is the perspective in which the relation of Kant's philosophy of geometry (based on the compositive form) to the discovery of the non-Euclidean geometries (based on the interpretative form) is most objectively understood.

Our reconstruction of the development of geometry puts before us the task to find (or to develop) also for the other forms of language of geometry, philosophical positions that would reflect them with comparable depth and precision as Poincaré reflected the integrative form of language of geometry. I am convinced that each form of language brings specific philosophical problems and requires a specific philosophical reflection. Besides a completion of the list of possible positions in the philosophy of geometry, such an analysis could reveal the vanity of some philosophical controversies. It seems that at least some controversies in the philosophy of mathematics are obscure because each side bases its arguments on a different form of language. Then arguments, which are acceptable and convincing from the viewpoint of one side, are considered by the other side weird and absurd.

2.3.3.2. The Form of Language and the Problem of Its Philosophical Reflection

Just as interesting as to compare different philosophical positions is to compare the philosophical reflection of a particular theory with actual mathematical practice. In this respect Poincaré's work offers a particularly interesting topic for analysis. Poincaré was both an outstanding mathematician and an author of important books on the philosophy of mathematics. Nevertheless, it is remarkable that these two areas of his work are only vaguely related. One could say that the mathematics which Poincaré reflected in his philosophical writings is very different from the areas of some of his most outstanding mathematical contributions. I am convinced that the notion of the form of language is able to explain this remarkable feature of Poincaré's work. Poincaré was one of the founders of algebraic topology, which brought the transition to the constitutive form of language in geometry. In the area of the philosophy of mathematics, on the other hand, Poincaré offered a penetrating epistemological reflection of the integrative form of language of geometry. It is precisely this difference of the form of language on which his scientific activity in geometry was based, and the area which formed the subject of his philosophical reflections, which illustrates the thesis that a language cannot express its own form. Poincaré offered a philosophical reflection of the integrative form of language, while the constitutive form of language, which he himself introduced into geometry, remained outside the realm of his philosophical reflections. Thus in analogy to Wittgenstein we could formulate a thesis that in a particular language it is possible to reflect only languages with a simpler form. It can be that it was precisely because Poincaré in his own mathematical work ascended above the integrative form of language that he was able to reflect this form in such a penetrating and precise manner.

If this phenomenon is not a mere historical accident, it could shed light onto one aspect of Kant's philosophy of geometry, which remained incomprehensible. It is possible to justify the view that in the area of epistemology Kant discovered the interpretative function of our mind. Thus according to Kant our mind not only passively registers the outside world, but it actively constructs our representation of it. Now interpretation is the fundamental aspect of the interpretative form of language, i.e., precisely of the form of language by means of which Gauss, Lobachevski, and Bolyai discovered the non-Euclidean systems. Therefore it is surprising that Kant used his discovery of the role of interpretation in our cognition for the articulation of a philosophical account of geometry, which can be interpreted as fully coherent with the compositive form of language.[29] Thus it seems that Kant in building his philosophy of geometry did not take advantage of

---

[29] According to Kant geometry is an *a priori* form of intuition of space. The role of this form is to introduce spatial order into the realm of sensory impressions. The Kantian form of intuition is thus a principle of coordination of sensuality. Thus Kant described the *coordinative form* (not of language, but rather of mind). Nevertheless, I would like to interpret Kant's philosophy as an articulation of the *compositive form*, because for Kant intuition was only one component of cognition and one of the goals of Kantian philosophy was to describe the components of cognition (sensitivity, intuition, understanding, reason) and to determine their mutual boundaries and limits. Even though in the Kantian philosophy we can find also stress of the active role of the mind in *interpreting* reality, in his philosophy of geometry Kant pronounced Euclidean geometry to be a priori and so he explicitly excluded the possibility of any other geometry than Euclidean. Thus in the field of geometry he did not exploit the possibilities opened by the discovery of the interpretative role of human mind. That is the reason why I would like to interpret Kant's philosophy of geometry on the basis of the coordinative form.

the possibilities offered by his own discovery of the role of interpretation in epistemology. It seems natural to bring this discrepancy in Kant's thought into connection with a similar discrepancy in the work of Poincaré.

In Kant's case the situation seems to be slightly more complicated. For Poincaré the constitutive form of language formed the basis of his mathematical work in geometry, while in his philosophical reflections he analyzed geometry based on the integrative form of language. Therefore here the situation is unequivocal. Poincaré philosophically reflected a form of language which preceded the form which he used in his own mathematical work. In Kant the situation is more complicated, because his creative act (the discovery of the integrative role of the mind) as well as his reflections (his philosophy of geometry) belong to the same area, namely to philosophy. Therefore in his work it is not possible to separate so easily the form of language that is the basis of his creative work from the form of language used in his philosophical reflections.

Maybe the different controversies that surround Kant's philosophy are a consequence of the fact that there are two forms of language present in his work. Those who wish to criticize Kant usually interpret Kant's views on the basis of the compositive form of language and then adduce some mathematical achievements (obtained on the basis of the interpretative form of language) that contradict Kant's views. On the other hand those who wish to defend Kant usually point to his discovery of the interpretative role of the mind in construction of our representations of the world. Thus according to the Neo-Kantians the discovery of non-Euclidean geometry has to be regarded as a vindication of Kant's philosophical views, rather than their refutation.

2.3.3.3. Kant's Philosophy and the Discovery of the Non-Euclidean Geometries

The views on Kant's philosophy that are scattered in the above passages can be united into a strategy of interpretation of Kant's philosophy of geometry. The discovery of the non-Euclidean geometries was accompanied by a fundamental linguistic innovation – the introduction of the interpretative distance in the perception of geometrical figures. As long as one is not willing to make this step, the non-Euclidean universe will not open up, and one will not be able to understand what the creators of the new geometries say. If we relativize Kant's philosophy and associate it with the corresponding form of language, it is possible to defend Kant's philosophy of geometry in a rational way. Kant's philosophy of geometry was not only a remarkable achievement of Kant's speculative mind, but it was a convincing philosophical reflection of geometry before the introduction of the interpretative form of language. This of course does not mean that we reject all critical arguments raised against Kant's philosophy of geometry. We reject only those which make use of the achievements obtained by means of the integrative form of language, for instance the arguments based on the discovery of non-Euclidean geometry. We reject them because these arguments cannot be formulated on the basis of the compositive form of language, i.e., the form that was dominant in Kant's own times.

2.3.3.4.   *The Great Number of Parallel Independent Discoveries of Non-Euclidean Geometry*

In the introduction to Chapter 2 we presented a quotation from Shafarevitch highlighting a remarkable aspect of the discovery of non-Euclidean geometries. Shafarevitch mentions the names of four mathematicians who "*independently, had come to the same results*" and in their works one "*sees the same designs as if drawn by a single hand*". Our approach makes it possible if not totally to explain then at least make this remarkable phenomenon understandable. As already mentioned on several occasions, there exist only a small number of possible forms of language. Therefore it is not so surprising that some mathematicians who were working at the beginning of the nineteenth century on the problem of the fifth postulate introduced the same form. The problem of the fifth postulate was unsolvable in the framework of the projective form of language, and so Gauss, Bolyai, Lobachevski, and Schweikart were forced to change the form of language. The small number of possible forms and the regularity of their changes make it probable that many or even all of them will make an innovation of the same kind – the transition to the interpretative form. Thus the designs mentioned by Shafarevitch were drawn not by a single hand, but by four different hands led by a single form of language.

Another problem is, why did it take fourteen centuries, since Proclus formulated the problem of the fifth postulate until the four mathematicians mentioned by Shafarevitch found independently from each other their solutions? I believe that also in this point our approach offers a plausible explanation. Before the interpretative form of language could be created, in the framework of which the non-Euclidean systems could be formulated, geometry had to pass the perspectivist and the projective forms, which of course required some time. So we see that the development of mathematics is not a result of purely contingent social conditions and random influences. There are a number of epistemological factors that are grounded in the changes of the forms of language in the development of mathematical theories, which play an important role in the history of mathematics. All questions that arose in connection with the discovery of non-Euclidean geometries cannot be reduced to the psychology of invention (i.e., to the question of who, when, and under what conditions came with the *bold guess* of the non-Euclidean systems, as Popper would put it) or to the sociology of the scientific community (as the problem would be formulated by Kuhn). The appearance of a particular theory is determined also by a number of epistemological factors that are related to the form of language in which the discovery takes place. To understand these factors can be important both from the point of view of the philosophy of mathematics as well as from the point of view of mathematics education.

2.3.4. The Changes of the Form of Language and the Development of Subjectivity

The relativizations that we found in the history of mathematics were accompanied by changes of the epistemic subject (which is one of the aspects of the form of language). The changes of the epistemic subject are interconnected in a remarkable way. Desargues, for instance, incorporated the epistemic subject of the Renaissance painters in the explicit form of the center of projection into the language. In this way he created the explicit projective subject. Lobachevski in his transition of the trigonometric formulas from the limit surface already used this projective subject, but, as the whole picture was impossible to draw within Euclidean geometry, he was forced to introduce a second, interpretative subject. Beltrami incorporated this interpretative subject in the form of explicit rules of translation between the external and the internal languages. In this way he created the explicit interpretative subject. Cayley disconnected the two languages, between which the translation took place and put the projective plane between them. To do this, he needed an additional structure of the language, based on a new kind of subject, the integrative subject. But for Cayley this subject was only implicit, and Klein found a way to incorporate it into the language in the form of transformation groups and so he created the explicit integrative subject. Riemann in his *Analysis Situs* made use of these transformation groups belonging to the integrative subject. But in order to be able to surpass the limitations which three-dimensional space put on geometry, Riemann introduced his cutting and pasting which were acts that belonged to the implicit constitutive subject. Poincaré in his combinatorial topology incorporated this implicit constitutive subject in an explicit way into language. Cantor discovered that all constitutive acts (i.e., acts performed by the constitutive subject) have a common structure – they consist in taking a system of objects as one whole. Set theory therefore uses an even more fundamental structure of subjectivity than that which is involved in the constitutive acts. Hilbert introduced this new conceptual subject into geometry, and just as in the other case he introduced it first in an implicit version. Tarski incorporated the implicit conceptual subject explicitly into language.

2.3.4.1. Dynamics of the Changes of the Epistemic Subject in Geometry

We see that the development of the form of language is not random. In this development, richer and richer structures were built into the language. First is incorporation of the subject having the form of the point of view, which is the basis of the *subjectiveness of the personal view*. Then follows incorporation of the subject having the form of interpretation, which is the basis of the *subjectiveness of meaning*. In the next step it is incorporation of the subject having the form of integration of all possible interpretations, which is the basis of the *subjectiveness of the possibilities of self-understanding*. An even further step is incorporation of the subject having the form of constitutive acts, which is the basis of the *subjectiveness of the possibilities of self-forming*. As a last step comes incorporation of the conceptual subject, which is the most fundamental structure of subjectivity that was hitherto used in geometry. We cannot deny that all these levels are parts of our own subjectivity. Everyone has his personal viewpoint, his own interpretation of reality, his own self-understanding, and his unique potentiality of possibilities. So we ourselves are the source from which geometry has taken the structures of its languages. In this development, deeper and deeper structures of our subjectivity were incorporated into the language. Therefore there is a simple answer to the question: *what is the development of geometry about?* It is about us, about our own subjectivity.

The epistemic subject of the language of geometry can be either implicit or explicit. We mean implicit or explicit in relation to language.

The implicit subject consists in the stance that we have to take in order to see, in two obviously convergent lines on the woodcut, two parallel sides of a ceiling; in some ordinary curves on a Euclidean plane non-Euclidean straight lines; in the Euclidean plane with its metric structure the projective plane stripped of the notion of distance and magnitudes of angles; or in a square *Klein bottle*. The implicit subject is not expressed in the language but, nevertheless, in the interpretation of the geometrical figures it is taken into account. In a sense the figures are drawn for it. The next step, which was taken by Desargues, Beltrami, Klein, Poincaré, and Tarski consisted in the incorporation of this implicit subject into the language. The subject became an explicit structure of the language. I suggest calling the epistemic ruptures that consist in the incorporation of deeper and deeper structures of the subject into the language *relativizations*. In their course, successive layers of our own subjectivity are reified in the language in the form of explicit objects, such as a point, a translation dictionary, or a symmetry group.

The relativization is closely related to objectivization. It can be said that each language allows us to objectivize only those relations which correspond to the structure of subjectivity, that is reified in that language. For instance if we wish to objectivize how a thing looks from different points of view, we need the Desargean language, in which the projective subject is reified. Then we can make use of the new possibilities offered by this language, which are based on the fact that the viewpoint is explicitly incorporated into the language in the form of a center of projection. Using this center of projection we can transform the point from which a particular thing is represented, and thus in an analytic way (i.e., making use only of the explicit rules of the syntax of the Desargean language) we can answer the posed question. Thus the structure of the subject, which is incorporated into the Desargean language, made it possible to determine in an objective way what a thing looks like from different points. We can therefore say that objectivity is constituted by relativization.

2.3.4.2.   A Deeper Understanding of the Development of Subjectivity Due to the Reconstruction of Algebra

When we transferred the notion of the form of language from geometry to algebra, we were led by epistemological analogies between these two disciplines. For instance Weber's factorization of a polynomial ring by an ideal is, from the epistemological point of view, parallel to Riemann's idea of cutting and pasting of surfaces in geometry. We called the form of language that is common to these two theories the constitutive form, because in both cases language takes over the task to constitute the objects of the theory.[30] As long as we considered the development of geometry and of algebra separately, the changes of the form of language seemed to be comprehensible. In each discipline we had to do with a sequence of clearly motivated shifts. But if we juxtapose the development of the form of language in geometry and in algebra, the question arises of what unifies these two developments. Why is it that in the development of geometry and of algebra, almost the same sequence of forms of language occurred? I would like to argue that the basis for these correlations and analogies in the development of the two mentioned disciplines is the evolution of the epistemic subject.

Let us look again at the scheme presented in Chapter 2.2.9 that summarizes the development of algebra. The left-hand column seems to represent the moments of birth of new kinds of subjectivity, the birth of a *new experience of ourselves*. In contrast to this, the forms of language in the right-hand column introduce a plurality into this experience of ourselves, which opens us to the *new experience of the other*. Thus the perspective form is about how *I* see the world, while by the help of the projective form of language I can understand how my world appears to *somebody else*. The coordinative form describes how *I* introduce order into my experience, while the compositive form allows me to bring the alternative orders introduced by *others* into harmony with my own. The interpretative form of language is about how *I* interpret my world, while the integrative form of language opens the possibility of understanding and comparing my interpretation of the world with those of *others*. The constitutive form of language reflects how *I* constitute myself, while the conceptual form of language opens the possibility of understanding alternative ways of self-constitution, those used by *others*. Thus the above described dynamics is one of encounters with deeper layers of understanding of ourselves as well as of encounters with others.

In a sense the diagram indicates that the road to a deeper layer of understanding of ourselves leads through the other. Thus in order to be able to encounter myself on the coordinative level as the source of order in the world and in this sense to discover my freedom as a subject

---

[30] It is possible to transfer our method of analysis from geometry also to the history of art (for details see Kvasz 2007). Of course the parallel between geometry and art is not as close as that between geometry and algebra, where often one mathematician contributed to both fields.

(which is perhaps the core of the *Cartesian experience of the self*), it is necessary first to encounter the otherness in the framework of the projective form of language. Only when I learn to see the world through the eyes of the other and thus free myself from the egocentric perspective, does it become possible to create a distance from the world, which is crucial for the coordinative form. Similarly, in order to be able to encounter myself on the interpretative level as the source of values and of interpretation of the world (which is perhaps the core of the *romantic experience of the self*), it is first necessary to encounter otherness in the framework of the compositive form of language. Only when I learn to tolerate the alternative orders of the world, and so free myself from the obviousness of the order I grew up in, does it becomes possible to discover the role and power of interpretative "distancing" from the world. Further, in order to be able to encounter myself on the constitutive level as the center of my existence (which is perhaps the core of the *existentialist experience of the self*), it is first necessary to encounter otherness in the framework of the integrative form of language. Only when I learn that all alternative interpretations of the world are basically equivalent (because they all are confronted with the same existential questions), and so free myself from the cultural framework in which my existence was veiled, am I prepared to encounter myself as a being that constitutes its existence.

Without the encounter with otherness we may be stuck on the surface of the self, on the outer layers of our subjectivity. But there is another danger. An encounter with the otherness of the other is made possible only by a deeper encounter with the self. Thus in order to be able to encounter the other in the projective form of language as a bearer of an alternative view of the world, and so to discover the fundamental plurality of worldviews (which may be the core of the *Mannerist fascination with otherness*), I must first encounter myself in the framework of the perspectivist form of language and learn to view the world from a fixed personal point of view. Similarly, in order to be able to encounter the other in the compositive form of language as a bearer of an alternative order of the world, and so discover the fundamental plurality of orders (which may be the core of the *Enlightenment's tolerance of otherness*), I must first encounter myself in the framework of the coordinative form of language and learn to subordinate my life to a fixed order. So too, in order to be able to encounter the other in the integrative form of language as a bearer of an alternative value system, and so to discover the fundamental plurality of evaluations (which may be the core of the *Positivist exploration of otherness*), I must first encounter myself in the framework of the interpretative form of language and learn to see these values as bases of interpretation.

This dynamics of the encountering of the deeper and deeper layers of our subjectivity can be seen as the source of the changes in geometry and in algebra. In this sense the history of algebra is an integral part of human culture, and its study may reveal many new things about ourselves.

2.3.5.   The Problem of Understanding Mathematical Concepts

If we introduce into the theory of mathematics education the pattern of change in the development of mathematics, which we discovered in the analysis of relativizations in mathematics, we obtain a tool for identification of the sources of misunderstanding in mathematics. Each mathematical concept is connected to a particular form of language in the framework of which the notion was introduced for the first time. So for instance the notion of substitution is connected to the projective form of language of algebra, while the notion of a group is connected to the integrative form. In the form in which it was first introduced, the notion has usually a rich semantic structure, the introduction of the notion is well motivated and is comprehensible. On the earlier levels of its development the form of language is too simple and so the language is not able to illustrate the different aspects of the particular notion in sufficient detail and precision. On the other hand, on later levels the language becomes too abstract and so it is not able to suitably motivate the introduction of the concept. Thus it seems that for every concept there is an optimal form of language, in the framework of which the concept is sufficiently specific and on the other hand having sufficiently rich content so that it can be correctly constituted (or constructed) in the minds of the students.

Mathematics has a tendency to empty its concepts of their original semantic content and to make them independent of the original motivation for which they were introduced. This emptying is not some freak behavior of the mathematicians, but a necessity. Mathematical creativity is often facilitated by the transfer of a notion, which was originally introduced in a particular area of mathematics, in a particular context, and for a particular purpose into a different area, a different context, and for different purposes. In order to make such transfers easier, since they are among the main sources of progress in mathematics, mathematicians naturally tend to decontextualize their concepts. They try to liberate concepts from their dependence on particular contexts, and the particular motivation of their introduction. When a concept becomes generally accepted and widely used by the mathematical community, it is to a large degree deprived of content. But what is perhaps even more important is that the new concept is also separated from the form of language to which it naturally belongs. A mathematical theory is usually formed of a set of concepts that have their origin in different forms of language; concepts that were introduced during different periods of the theory's development. When mathematicians present their theories to students, they usually present them in a decontextualized form. Then it depends on the maturity of the student which part of the theory he or she is able to recontextualize and integrate into his or her own semantic network. Of course, the more advanced form of language the student reaches in his or her personal development, the greater is his or her chance to understand what the teacher presents to him or her. On the other hand students who in their personal development have reached only the less advanced forms of language will not be able to follow the teacher's expositions.

I am convinced that the reconstruction of mathematics presented in this part of the present book offers the teacher an effective tool for enhancing his teaching. For each problematic notion or theorem he can ask, to which form of language does it naturally belong. When the theory is too inhomogeneous and contains material belonging to different forms of language (which in mathematics is more a rule than an exception), the epistemological reconstruction makes it possible to order the different parts of the theory according to the growing complexity of its form of language, which will make his or her exposition more comprehensible. When his or her students encounter difficulties in following his or her exposition, our theory makes it possible to localize the probable causes of these difficulties. In the majority of cases the misunderstandings and misconceptions in teaching and learning mathematics have their roots in the differences between the form of language used by the teacher in the exposition of his subject and the form of language on the basis of which the student tries to comprehend it. Usually the teacher is not aware of the differences between the particular forms of language and so he presents, during one lecture, notions and propositions that belong to three or even more forms. Using the method of reconstruction presented in this chapter it is possible to clarify, for each concept or theorem that causes difficulties for students, to which form of language it belongs. The next step is to identify the highest form of language which the students have mastered to a sufficient degree, and then to find a strategy for connecting the form of language that is natural to the student with the one that causes him or her difficulties. In some cases it is easy to find such a strategy; in others, especially when the distance between the two forms is greater, it is necessary to proceed piecemeal and to introduce one or more intermediate steps. The attentive reader might have noticed a close connection between our approach and constructivism in mathematics education that is based on the Piagetian approach. Perhaps the only difference is that we base our approach not so strongly on psychology, but rather on the epistemological reconstruction of the development of mathematics.

2.3.6.   A Gap of Two Centuries in the Curricula

A remarkable aspect of the teaching of many mathematical disciplines is the gap that separates the high school and the university curricula. This phenomenon is perhaps most salient in mathematical analysis, where high school stops somewhere near the end of the seventeenth century (introducing the notion of co-ordinates, some elementary properties of functions of one variable, and the notions of derivative and integral). On the other hand, the university curricula start in most countries with the $\varepsilon - \delta$ *analysis*, i.e., somewhere in the middle of the nineteenth century. In this way at least two centuries of development are excluded from the curricula. A similar gap can be found also in algebra. Here too high school ends somewhere near the end of the seventeenth century (manipulation with polynomials and other algebraic expressions). University courses in algebra usually start with the notion of a *group* (a field or a vector space). These notions are usually introduced in an abstract axiomatic form, so that the university requires the student to pass from the coordinative form of language straight to the constitutive one. Just as in mathematical analysis, so also in algebra there is a gap caused by skipping the compositive and the interpretative forms of language. An analogical gap exists also in geometry, where the university course starts with *topology*. Topology is based on the constitutive form of language. The student does not have much chance to comprehend this form if he had not understood the interpretative and the integrative form earlier. Just as in analysis or group theory, so also in topology the student can memorize the subject and understand the particular proofs and constructions. Nevertheless, the overall meaning of topology remains hidden for him.

It is probable that it is this gap in the curricula that is responsible for the shock that many students experience in the first year of their university studies. Our reconstruction of the development of algebra and of geometry has shown that the consecutive forms of language that emerge in the development of a mathematical discipline are linked to each other. Therefore it is not possible to understand a topic that is based on the constitutive form of language without mastering the interpretative and integrative forms. Every transition from one form of language to the next, taken separately, is comprehensible and so does not present any substantial educational problem. But if we skip two or three intermediate levels, as happens between high school and the university, the transition becomes incomprehensible and the students have no chance to follow.

CHAPTER 3

Re-Formulations as a Third Pattern of Change in Mathematics

The changes that were analyzed in the previous two chapters were of a global nature. Each of them consisted in the rebuilding of the syntactic or semantic structure of wide areas of mathematics. Therefore they happened rarely. They were not the fruit of the work of one mathematician but rather the result of the work of a whole series of them. The resulting language that emerges from their collective effort is usually a compromise that unites the innovations originating from many different authors. In contrast to this, the changes that will be analyzed in the present chapter are of a local nature. Usually they are related to a single definition, theorem, proof, or axiom. They happen often and form the content of the everyday work of mathematicians. These changes consist in the reformulation of a problem, a definition, a proposition, or an axiom and therefore I suggest calling them *re-formulations*. The hyphen in the name means that it is not an arbitrary reformulation but one which introduced a change important from the epistemological point of view. We can formulate a problem, a definition, a proposition, or an axiom in a particular language in many different ways. In some contexts these formulations may be equivalent and we can use them as synonyms. We will consider a transition from one formulation to another to be a *re-formulation* only if there is a mathematically relevant context, in which these two formulations enter into fundamentally different connections, shed different light on their surroundings, and direct the lines of thought on different paths.

An example that illustrates the notion of a re-formulation is Euclid's fifth postulate. Euclid formulated it around 300 BC. as:

> "That, if a straight line falling on two straight lines make the interior angles on the same side less than two right angles, the two straight lines, if produced indefinitely, meet on the side on which are the angles less than the two right angles." (Euclid, p. 202)

An alternative formulation of the fifth postulate is from John Playfair from 1795:

> "Given a line l and a point P not on it we can draw one and only one line through P which does not meet l." (Gray 1979, p. 87)

The historian Jeremy Gray commented on Playfair's formulation saying:

> "Its great attraction is that it can readily be reformulated to suggest non-Euclidean geometries by denying either the existence or uniqueness of parallels." (Gray 1979, p. 34)

In the context of Euclidean geometry both formulations, Euclid's and Playfair's, are mathematically more or less equivalent. Nevertheless, there is a context; in this case it is the context of the non-Euclidean geometries, in which Playfair's formulation of the fifth postulate sheds more light on the possible alternatives of Euclidean geometry. Therefore I suggest considering Playfair's formulation to be a *re-formulation* of Euclid's original formulation.

Another possible way to formulate the fifth postulate would be:

> "To construct a square with any center and side."

This formulation is in consonance with the spirit of the other postulates (it has an analogous wording as the third postulate, which says: "To describe a circle with any centre and distance"), and is of course much shorter than Euclid's (and even than Playfair's) formulation. We have to be grateful to Euclid that he chose his complicated formulation and not this rather evident proposition. In the context of Euclidean geometry all three formulations are equivalent; each one is sufficient to secure the Euclidean nature of the plane. Nevertheless, despite their logical equivalence there are fundamental epistemological differences between them. Had Euclid chosen the formulation with the squares, probably nobody would have ever doubted it and the history of geometry would have taken a different course. It was Euclid's cumbersome formulation of the fifth postulate that led Proclus to the conviction that this proposition is not a postulate (i.e., a self-evident truth), but rather a theorem, i.e., something that has to be proven. The efforts of generations of mathematicians to find a proof of the fifth postulate become incomprehensible if we replace the postulate by Playfair's formulation, and they would become even an absurdity, if we used the formulation with squares.

In the case of Playfair we will speak about a re-formulation of the fifth postulate. But it is important to realize that in the eighteenth century, when he introduced this re-formulation, the context of the non-Euclidean geometries did not exist, and so Playfair could not foresee the fruitfulness of his move. Only thanks to the discovery of the non-Euclidean systems did it turn out that, from Playfair's formulation, there is a direct path to the non-Euclidean worlds. This shows that just as in the case of re-codings and relativizations, so also in the case of re-formulations we mean an objective change in the language of mathematics, which is in a sense independent of whether the authors of the change were or were not aware of its importance. The case of the fifth postulate indicates that sometimes a cumbersome and awkward formulation, like that used by Euclid, can be stimulating and fruitful for further development of the discipline. This case also indicates that, despite being changes of a rather small scale, re-formulations can have a fundamental impact on the course of the development of mathematics.

Re-formulations are most often encountered in the form of multiple independent discoveries. As examples we can take the differences in the formulation of the differential and integral calculus by Newton and Leibniz; the differences in the formulation of the geometrical interpretation of the complex numbers by Wessel, Argand, Warren, and Gauss; or the differences in the formulation of the non-Euclidean geometry by Gauss, Bolyai, and Lobachevski. Of course, the first of these discoveries was a re-coding, while the other two were relativizations. Nevertheless, each of the discoverers of a new theory brings his own formulation of it; a formulation which has some advantages in some areas and disadvantages in others. Therefore what becomes the standard formulation and enters into the textbooks is usually neither one of the original formulations proposed by the discoverers, but a re-formulation, which tries to unite the advantages of each of them while omitting all the insufficiencies.

The study of re-formulations offers rich material for epistemological analyses. The study of the different re-formulations that a theory undergoes on its road from discovery to standardization as well as the comparison of the different alternative formulations, in which the theory was invented, is doubtlessly important from the historical point of view. In contrast to the other two chapters of the present book, in the case of re-formulations I could not find any regular developmental pattern. Therefore I will not present any "historical description of re-formulations" and I will not attempt any philosophical or educational reflection of them. I will introduce only a few representative examples. For a better overview I decided to divide these examples into three classes: re-formulations in *concept-formation*, re-formulations in *problem-solving*, and re-formulations in *theory-building*.

3.1.   Re-Formulations and Concept-Formation

Re-formulations play an important role in the process of the formation of mathematical concepts and the search for their appropriate definitions. A classic in this area is the book *Proofs and Refutations* by Imre Lakatos. The book contains a reconstruction of the process that started with the proof of Euler's theorem on polyhedra and turned into the search for the definition of a polyhedron. Lakatos succeeded in capturing an important aspect of the creative process in mathematics and to describe the techniques by means of which mathematicians confront the potential counterexamples to their theories. But besides its interesting content, the success of *Proofs and Refutations* was also determined by its remarkable form. The book is written in the form of a dialogue, which takes place in a classroom. But it is not an ordinary classroom. Lakatos has brought together the greatest mathematicians of the past who contributed to the theory of polyhedra – from Cauchy and Lhuilier to Abel and Poincaré. It is exciting to imagine what would happen if all the participants in a scientific debate which lasted over two centuries could discuss the problems together. How would Cauchy react to the counterexamples to his theorem? Would he accept Poincaré's topological proof? So, the very idea of such an imaginary dialogue is interesting. But Lakatos achieved even more. He succeeded in distilling from the history some basic patterns of thought which can be found in many other areas of mathematics. These are his famous *monster* *barring, exception barring,*and *lemma incorporation*. We will present them briefly.

The discussion in the classroom is of Euler's theorem which says that, for all polyhedra, the number of vertices V, number of edges E, and number of faces F fulfil a simple relation: $V - E + F = 2$. After the teacher has presented a proof – actually, the classical proof stemming from Cauchy – some counterexamples appear. I will not present all the counterexamples discussed in the book; I select just a few of them here, and present the basic ideas.

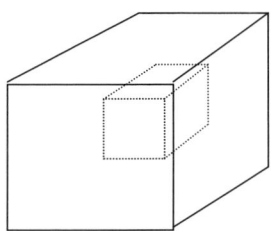

Example 1: A cube, which has inside of it an empty hole in the form of a smaller cube. It easy to see that, in this case, $V - E + F = 4$, and not 2, as it should, according to the theorem.

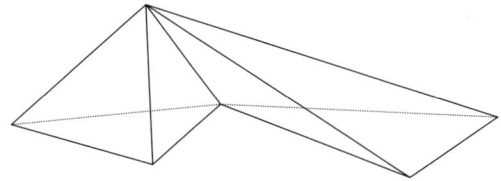

Example 2: Two tetrahedra which have one edge in common. Here we have $V - E + F = 3$.

Surely these examples represent, so to speak, principles of how to construct some fairly strange objects! In this sense, it is not difficult to imagine an object having many holes, or to build a whole chain of different polyhedra in which any two neighbors have a common edge. Now the question is how to deal with these objects, which contradict Euler's theorem.

The first strategy described by Lakatos is that of *monster-barring*. These strange objects are surely not what we have in mind when we speak of polyhedra. They are monsters, and we should not allow them to enter into our considerations. They are of no theoretical interest, and no normal mathematician would ever think of them as polyhedra. The second strategy Lakatos calls *exception-barring*. According to it, we admit that these objects are genuine polyhedra, and therefore real counterexamples to the theorem. The theorem, as it was stated originally, does not hold. It is not so general as we have formerly thought. We have to restrict our theorem in such a way that all these exceptions would fall outside its domain. It is obvious that all the examples mentioned above are not convex. Thus, if we restrict ourselves to convex polyhedra, the theorem is saved. The third strategy described by Lakatos is that of *lemma incorporation*. In both previously described cases, we have not learned much new from our new objects. In *monster-barring*, we just ignore them and state the theorem more generally than it really holds. On the other hand, in the *exception-barring*, we restricted the theorem too much. We should not restrict the original theorem, but should, rather, try to find a more general one, which would also include the strange objects. Only in this way can we learn something really new. Thus, we should try to understand in what way a common edge or a hole changes the resulting statement of the theorem, and then find a way of incorporating them into the theorem. Our task is not to find a safe ground upon which the theorem holds (as in *exception-barring*). We have first to understand what new things we can learn from these objects.

Koetsier in his book *Lakatos' Philosophy of Mathematics, A Historical Approach* (Koetsier, 1991) compared Lakatos' reconstruction of the development of the theory of polyhedra from *Proofs and Refutations* with the actual history. He gives the following account:

> "There is some resemblance between Lakatos's reconstruction concerning the formula of Euler for polyhedra and the real history. ... Yet the rational construction deviates considerably from the chronological order in which things actually happened. ... As far as chronology is concerned, there is not much in common between dialogue and real history. There is no doubt that *Proofs and Refutations* contains a highly counterfactual rational reconstruction.." (Koetsier 1991, p. 42)

Despite this, Koetsier's overall judgment of Lakatos' book is positive: "*Proofs and Refutations* is mainly convincing because it shows recognizable mathematical behavior" (Koetsier 1991, p. 44).

But even if we admit that Lakatos' rational reconstruction deviates from actual history, the real question is whether it offers a true picture of mathematical behavior itself? Doubts in this regard were expressed by Corfield (Corfield 1997), who argued that the counterexamples do not play as important a role in mathematics as Lakatos gives them in his rational reconstruction. He cites the example of Poincaré's *Analysis situs*, the proofs of which were changed to a large extent; there was, nevertheless, only one counterexample to a single lemma in all 300 pages of the text (Corfield 1997, p. 108). Thus, the reformulations of Poincaré's proofs and theorems were not the result of the discovery of counterexamples. That means that there are other patterns of mathematical behavior, which were omitted in Lakatos's reconstruction.

One such clearly recognizable pattern can be found in Koetsier's book. We have in mind the proof of the interchangeability theorem for partial differentiation by H. A. Schwarz. Schwarz first stated the theorem of interchangeability with six conditions, proved it, and then attempted to drop as many of the conditions as possible. He succeeded in dropping three of the six conditions, and ended with a much stronger theorem than the one which he initially proved (Koetsier 1991, pp. 268–271). I suggest calling this method *lemma-exclusion*, and to consider it a counterpart to *lemma incorporation*. If we incorporate this fourth method into Lakatos' theory, we get a more balanced schema. According to this schema, we have two possible reactions on the appearance of a counterexample. One is to ignore the counterexample as a monster, the other is to consider it as an exception and restrict the theorem to safe ground. Nevertheless, *monster-barring* states the theorem more generally than it really is. On the other hand, *exception-barring* often restricts the theorem too strongly (for instance, Beta in *Proofs and Refutations* p. 28, who restricted Euler's theorem to convex polyhedra, or H. A. Schwarz, who in his first theorem restricted the interchangeability theorem only to functions fulfilling all the six conditions). After some time, these first reactions are overcome. The *monster-barring* method is followed by *lemma-incorporation*, where the aim of having the theorem to be as general as possible is still preserved, but the counterexamples are no longer ignored. On the other hand, *exception-barring* is followed by *lemma-exclusion* (as the case of H. A. Schwarz suggests), where the aim of being constantly on safe ground is preserved, but the too-strong restrictions of the domain of the theorem are gradually weakened. In the ideal case, *lemma-incorporation* and *lemma-exclusion* meet each other when the necessary and sufficient conditions of the theorem are found.

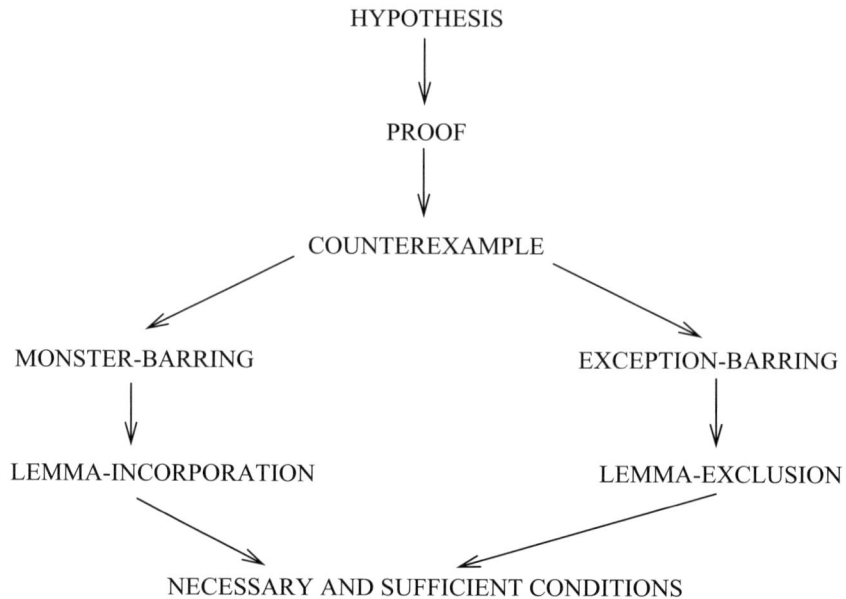

We see that re-formulations are not restricted to problem-solving. Also after a problem has been successfully solved or a theorem has been proven, it makes sense to try to reformulate the theorem. In the process of re-formulation of theorems and their proofs, many important mathematical concepts were created. Also several definitions that can be found in mathematical textbooks are the result of re-formulations, which were made necessary by the appearance of counterexamples or were stimulated by the effort to give a theorem the possible largest generality.

3.2.   Re-Formulations and Problem-Solving

Another area closely related to re-formulations is the solution of mathematical problems. This area has a rich literature starting with the classical works of George Polya. His *How to solve it?* was published in 1945, and more than a million copies have been sold. As one of the heuristics in problem-solving Polya mentioned re-formulation: "*If you cannot solve the proposed problem, try to solve first some related problem*" (Polya 1945, p. 23). Also his other two books, *Mathemat-*

ics and Plausible Reasoning (Polya 1954) and *Mathematical discovery* (Polya 1962) contain many examples of re-formulations in problem-solving. As an illustration of the heuristic force of re-formulations we can take the problem of determining the sum of the reciprocal values of the squares of all natural numbers

$$1 + \frac{1}{4} + \frac{1}{9} + \frac{1}{16} + \frac{1}{25} + \frac{1}{36} + \frac{1}{49} + \frac{1}{64} + \frac{1}{81} + \cdots .$$

This problem seized the attention of Euler, who first calculated the sum of this series to six decimal places and obtained the value 1.644934. Unfortunately he was not able to find any regularity in this decimal expansion and the number itself did not remind him of any known number. Several years later Euler succeeded in determining the value of the above sum by means of a remarkable re-formulation.

It is known that if a polynomial $a_0 + a_1 x + a_2 x^2 + \ldots + a_n x^n$ has n different real roots $\alpha_1, \alpha_2, \ldots, \alpha_n$, none of which equals 0, then the polynomial can be written as

$$a_0 + a_1 x + \ldots + a_n x^n = a_0 \cdot \left(1 - \frac{x}{\alpha_1}\right) \cdot \ldots \cdot \left(1 - \frac{x}{\alpha_n}\right).$$

Although this is not the common form we use to write a polynomial, it presents no difficulties. We have written the polynomial as the product of expressions of the form $\left(1 - \frac{x}{\alpha}\right)$, and we have taken one such expression for each root of the polynomial. Therefore the fact that both expressions represent the same polynomial can be inferred from the fact that on both sides we have a polynomial of the n-th degree, which has the roots $\alpha_1, \alpha_2, \ldots, \alpha_n$ and for $x = 0$ it takes the value a_0. It is known from elementary algebra that a polynomial is uniquely determined by these conditions.

From the above identity it is not difficult to deduce that

$$a_1 = -a_0 \cdot \left(\frac{1}{\alpha_1} + \frac{1}{\alpha_2} + \cdots + \frac{1}{\alpha_n}\right).$$

It is sufficient to compare the coefficients that stand by the first power of x in both expressions of the polynomial. As these expressions express the same polynomial, these coefficients must be equal.

Let us now turn to the slightly more complicated case of the polynomial $b_0 - b_1 x^2 + b_2 x^4 - \ldots + (-1)^n b_n x^{2n}$, which differs from the previous one in that it has only even powers of the unknown. Let us suppose that we know all the $2n$ roots of this polynomial. Let us further suppose that all of these roots are mutually different real numbers $\beta_1, -\beta_1, \beta_2, -\beta_2, \cdots, \beta_n, -\beta_n$. For every root β the list contains also the root $-\beta$, because the polynomial has only even powers of the unknown, and even powers of β and $-\beta$ give the same value. Thus when β is a root of the polynomial, $-\beta$ is a root as well. In an analogous way to the above we obtain

$$b_0 - b_1 x^2 + b_2 x^4 - \ldots + (-1)^n b_n x^{2n}$$

$$= b_0 \cdot \left(1 - \frac{x^2}{\beta_1^2}\right) \cdot \ldots \cdot \left(1 - \frac{x^2}{\beta_n^2}\right). \quad (3.1)$$

From this identity we can obtain, by comparing the coefficient by x^2, that

$$b_1 = b_0 \left(\frac{1}{\beta_1^2} + \frac{1}{\beta_2^2} + \cdots + \frac{1}{\beta_n^2}\right).$$

All this is rather elementary and not very exciting. But now comes the crucial step of the whole derivation. Let us imagine the infinite series for the sine function, which was familiar to Euler:

$$\sin(x) = x - \frac{x^3}{6} + \frac{x^5}{120} - \frac{x^7}{5040} - + \cdots.$$

Euler's idea was to view this series as a "polynomial" of an infinite degree. Then the numbers $0, \pi, -\pi, 2\pi, -2\pi, 3\pi, -3\pi, \ldots$, in which the function $\sin(x)$ has the value 0, become "roots" of this "polynomial". In order to get rid of the "root" zero (and thus to be able to use the expression (3.1)), Euler divided the series for $\sin(x)$ by x. In this way he obtained another "polynomial" of an infinite degree:

$$\frac{\sin(x)}{x} = 1 - \frac{x^2}{6} + \frac{x^4}{120} - \frac{x^6}{5040} - + \cdots$$

which has only nonzero "roots" $\pi, -\pi, 2\pi, -2\pi, \ldots$. Therefore in analogy to (3.1) Euler wrote:

$$1 - \frac{x^2}{6} + \frac{x^4}{120} - \frac{x^6}{5040} - + \cdots =$$

$$= \left(1 - \frac{x^2}{\pi^2}\right) \cdot \left(1 - \frac{x^2}{4\pi^2}\right) \cdot \left(1 - \frac{x^2}{9\pi^2}\right) \cdots. \quad (3.2)$$

From here we obtain for b_1 (the coefficient by $-x^2$ on the left-hand side as well as on the right-hand side of the identity (3.2)) the expression

$$\frac{1}{6} = \frac{1}{\pi^2} + \frac{1}{4\pi^2} + \frac{1}{9\pi^2} + \frac{1}{16\pi^2} + \frac{1}{25\pi^2} + \cdots .$$

Multiplying both sides by π^2 we obtain the surprising result

$$\frac{\pi^2}{6} = 1 + \frac{1}{4} + \frac{1}{9} + \frac{1}{16} + \frac{1}{25} + \frac{1}{36} + \frac{1}{49} + \cdots .$$

This is a fascinating result because it shows that even if geometry did not exist and we knew only the natural numbers, we could arrive at the number π. The fact that we are used to consider the number π as belonging to geometry (the area of the unit circle) is only a historical contingency. Geometry was that branch of mathematics where we first came across this number. Euler's discovery shows that the number π belongs with the same right to arithmetic. In his letter of March 13, 1736 Euler mentioned that $\frac{\pi^2}{6} = 1.6449340668482264364$. If we realize that in Euler's times there were no computers, this result represents both Euler's ingenuity and patience. It contains in a condensed form long hours of calculations.

This result is the consummation of a fascinating chain of discoveries at the beginning of which stood a simple *re-formulation*; Euler's idea to view the infinite series expressing the function $\sin(x)$ as a polynomial and to represent this polynomial as a product of quadratic factors. Of course not all re-formulations lead to such spectacular results, but Euler's examples indicate most sharply the importance of re-formulation in problem-solving. It would be possible to adduce hundreds of similar fascinating examples, where a simple re-formulation of the problem opened the road to its solution. I consider it better to refer the reader to the voluminous literature on this subject, as the *Triumph der Mathematik* (Dörrie 1932) or *Proofs from the Book* (Aigner and Ziegler 1998).

3.3. Re-Formulations and Theory-Building

Besides problem-solving and concept-formation, a further area where re-formulations occur is the area of axiomatisation of mathematical theories. When a sufficient number of related problems have been solved and the analysis of the solution methods have given rise to a set of important concepts, the need for the unification of the new area appears. In modern mathematics unification is often achieved through axiomatisation. It is sufficient to mention Frege's axiomatisation of the predicate calculus, Zermelo's axiomatisation of set theory, Peano's axiomatisation of arithmetic, or Kolmogorov's axiomatisation of probability theory. Cases like Frege's, where the first axiomatic system was already complete, are rare. More often the definitive axiomatisation is achieved only gradually, through a series of re-formulations. It can also happen that for one theory several alternative axiomatisations are offered (as for instance Zermelo's and von Neumann's systems for set theory). Some axioms can meet with opposition and alternative axioms may be invented in order to replace it. This shows that re-formulations play an important role in the creation of a suitable axiomatic system for a mathematical discipline. The fine differences between alternative formulations of a particular axiom are crucial for the mathematical and meta-mathematical properties of the theory. To highlight the importance of re-formulations for theory-building, just as in the case of problem-solving and concept-formation, I will present a single example – Zermelo's axiomatisation of set theory. In his *Untersuchungen über die Grundlagen der Mengenlehre*[1] from 1908 Zermelo introduced the following seven axioms:

Axiom I.   (*Axiom of extensionality.*) If every element of a set M is also an element of N and vice versa, if therefore both, $M \subseteq N$ and $N \subseteq M$, then always $M = N$; or more briefly: Every set is determined by its elements.

Axiom II.   (*Axiom of elementary sets.*) There exists a (fictitious) set, the *null set*, 0, that contains no element at all. If a is any object of the domain, there exists a set $\{a\}$ containing a and only a as element; if a and b are any two objects of the domain, there always exists a set $\{a, b\}$ containing as elements a and b but no object x distinct from both.

Axiom III.   (*Axiom of separation.*) Whenever the propositional function $E(x)$ is definite for all elements of a set M, M possesses a subset

---

[1]   This and the following classical papers on set theory are quoted from the anthology *From Frege to Gödel* (van Heijenoort 1967).

M_E containing as elements precisely those elements x of M for which $E(x)$ is true.

Axiom IV. (*Axiom of the power set.*) To every set T there corresponds another set UT, the *power set* of T, that contains as elements precisely all subsets of T.

Axiom V. (*Axiom of the union.*) To every set T there corresponds another set ST, the *union* of T, that contains as elements precisely all elements of the elements of T.

Axiom VI. (*Axiom of choice.*) If T is a set whose elements all are sets that are different from 0 and mutually disjoint, its union ST includes at least one subset S_1 having one and only one element in common with each element of T.

Axiom VII. (*Axiom of infinity.*) There exists in the domain at least one set Z that contains the null set as an element and is so constituted that to each of its elements a there corresponds a further element of the form $\{a\}$, in other words, that with each of its elements a it also contains the corresponding set $\{a\}$ as an element.

We can see that the axiom of choice was present in Zermelo's first axiomatisation. An interesting re-formulation of axiom III was offered by Fraenkel in his paper *Der Begriff "definit" und die Unabhängigkeit des Auswahlsaxioms* from 1922. Fraenkel criticized Zermelo for the use of the imprecise notion "*definit*" in the formulation of the axiom of separation. In order to replace it, Fraenkel introduced the notion of a *function*, which he characterized as a rule of the following kind:

> "an object $\phi(x)$ shall be formed from a ('variable') object x that can range over the elements of a set, and possibly from further given ('constant') objects, by means of a prescribed application (repeated only a finite number of times, of course, and denoted by ϕ) of Axioms II–VI. For example, $\phi(x) = \{\{\{x\}, \{0\}\}, Ux + \{\{0\}\}\}$." (Fraenkel)

The axiom of separation then acquired the form:

Axiom III.   (*Axiom of separation.*) If a set M is given, as well as, in a definite order, two functions ϕ and ψ, then M possesses a subset M_E (or a subset $M_{E'}$) containing as elements all the elements x of M for which $\phi(x)$ is an element of $\psi(x)$ (or for which $\phi(x)$ is not an element of $\psi(x)$), and no others.

Another re-formulation of Zermelo's system stems from Skolem, who in his *Einige Bemerkungen zur axiomatischen Begründung der Mengenlehre* from 1922 suggested to add to Zermelo's axioms a further one, which is called the axiom of substitution. He formulated it as follows:

Axiom VIII.   (*Axiom of substitution.*) Let U be a definite proposition that holds for certain pairs (a, b) in the domain B; assume, further, that for every a there exists at most one b such that U is true. Then, as a ranges over the elements of a set M_a, b ranges over all elements of a set M_b.

The last axiom that is usually introduced among the axioms of set theory is the axiom of foundation. It was introduced by John von Neumann in his *Eine Axiomatisierung der Mengenlehre* from 1925. As von Neumann's paper belongs to another approach to set theory (in which besides sets also classes exist), and von Neumann formulated his axiom not in terms of sets but of functions, his original formulation of the particular axiom is more difficult to comprehend. Therefore I will quote it in the form in which Fraenkel and Bar–Hillel introduce it:

Axiom IX.   (*Axiom of foundation.*) Every nonempty set a contains such an element b, that a and b have no common elements.

We see that the axiom system that is nowadays used as the standard formulation of set theory was created during a period of 17 years on the basis of the contributions of four authors. Alongside the process of standardization of the axiomatic system of set theory, a heated debate took place concerning the axiom of choice. Several alternatives to the axiom of choice were proposed, and even an alternative formulation of the whole Cantorian set theory appeared in the form of *Alternative set theory* of Vopenka (see Vopenka 1979). The discussion of this fascinating area of research would, unfortunately, lead us too far from the topic of the present book.

CHAPTER 4

Mathematics and Change

We have reached the end of our exposition of the patterns of change in the development of mathematics. In the closing chapter I would like to discuss some interpretations of the development of mathematics. I will focus on the approaches based on the works of Kuhn, Lakatos, and Piaget and I will compare them with the approach described in the present book. A characteristic feature of the latter approach is that it emphasizes the historicity of the language of mathematics – it studies mathematical knowledge in relation to the historically existing expressive tools of the formal language of mathematics. Thus in a sense it is a compromise between the *analytical approach* in the philosophy of mathematics, which stresses the importance of language and logic, but understands them in an unhistorical way; and the *historical approach*, which usually stresses the historical nature of knowledge, but studies the historical growth of knowledge using sociological or psychological means. The approach developed in this book is a middle position between these two approaches. On the one side it accepts that mathematical knowledge is in a fundamental sense historical, but on the other hand it tries to grasp this historicity using the means of the analytical tradition.

4.1. Revolutions in Mathematics (Kuhn)

The works of three historians of mathematics sparked a vivid discussion of the possibility of using Kuhn's theory of scientific revolutions in the description of the development of mathematics. Michael Crowe in his paper *Ten "laws" concerning patterns of change in the history of mathematics* formulated an unequivocally negative stance, according to which "Revolutions never occur in mathematics." (Crowe 1975, p. 165). A year later in the paper of Herbert Mehrtens, *T. S. Kuhn's theories and mathematics: a discussion paper on the "new historiography" of mathematics*, a moderate approach was taken towards the possibility of using Kuhn's theory in the history of mathematics. Mehrtens argues that some notions (such as scientific community, anomalies, and normal science) have an explanatory value and offer a tool for the historical study of mathematics, while others (such as revolution, incommensurability, and crisis) are without any explanatory value in mathematics and direct the discussion into nonproductive quarrels. Nevertheless, according to Mehrtens, Kuhn's theory as a whole cannot be applied to the history of mathematics. In 1984 appeared the paper *Conceptual revolutions and the history of mathematics: two studies in the growth of knowledge* by Joseph Dauben, where two examples were introduced – the discovery of incommensurable magnitudes and the creation of set theory – to which, according to the author the theory of Kuhn is fully applicable. These three papers were reprinted in the book *Revolutions in Mathematics* (Gillies 1992). Besides the mentioned three papers the book contains nine newly written papers, in which different historians of mathematics analyze crucial moments in the development of mathematics and discuss the question whether they were revolutions or not. In connection with this discussion several questions appear.

4.1.1. What Kind of Changes Should we Consider Revolutionary?

If we distinguish three kinds of change in the development of mathematics (re-codings, relativizations, re-formulations), then there naturally arises the question: what kind of changes can be viewed as revolutions? It is interesting that some of the examples discussed in the book *Revolutions in mathematics* were re-codings (the creation of set theory (Dauben 1984), the discovery of analytic geometry (Mancosu 1992), the discovery of the calculus (Grosholz, 1992), and the discovery of the predicate calculus (Gillies 1992a)) while others (the legitimization of the calculus (Giorello 1992), the discovery of non-Euclidean geometry (Zheng 1992), the changes in ontology (Gray 1992), and the discovery of non-standard analysis (Dauben 1992)) were relativizations. Thus, in principle, there are three possibilities of defining the notion of a revolution in mathematics.

The first possibility (espoused by Crowe) is to see re-codings as well as relativizations as conservative changes, i.e., changes that are not revolutionary. The analyses in the present book support this view, because they show a high level of regularity both in re-codings as well as in relativizations. Because of the regular pattern of these changes it seems not to be appropriate to call them revolutionary. The second possibility would be to consider re-codings as revolutionary (because they change the universe of mathematics), while relativizations would be considered as conservative changes. This could be the position of Dauben, or at least this can be inferred from the two examples that he discussed, as both of them were re-codings. A third possibility would be to consider the notion of revolutions in mathematics to apply to re-codings as well as to relativizations. This position could be close to the views of Gillies who in the introduction to the *Revolutions in Mathematics* suggested distinguishing two meanings of the word revolutions (and so to reconcile the views of Crowe and Dauben as not contradicting each other, but rather speaking about different things). While Gillies used political vocabulary in his distinction, our classification offers him the possibility to define the two kinds of revolutions in an epistemological way – as re-codings and relativizations.

4.1.2. The Distinction between Revolutions and Epistemological Ruptures

Another set of problems that arises from the confrontation of our approach with Kuhn's theory is that they are built on different assumptions, which should be made more explicit. Maybe it would be wise to retain for the notion of *scientific revolution* its original sociological content. Instead of trying to introduce different kinds of revolutions, we could introduce for the changes that were analyzed in this book a new term, for instance *epistemic ruptures*. These changes are discontinuities (i.e., ruptures), and they were identified on the basis of the analysis of the language of mathematical theories and their different epistemological aspects (as logical or expressive power in the case of re-codings; or the nature of the epistemic subject and the horizon in the case of relativizations). The distinction between scientific revolutions and epistemic ruptures opens new possibilities for the study of science.

To each scientific revolution there corresponds an epistemic rupture that summarizes the changes of the syntax and semantics of the scientific language that occurred during the revolution. Nevertheless, not every epistemic rupture of a particular kind turns into a scientific revolution, as can be seen by the example of synthetic geometry. In Section 2.1 we analyzed six epistemic ruptures that occurred in the history of synthetic geometry. But in the literature only one of them is discussed as an example of a revolution in mathematics, namely the discovery of non-Euclidean geometry. So the distinction between scientific revolutions and epistemic ruptures makes it possible to formulate a question of great epistemological importance: *which epistemic ruptures become scientific revolutions?* It is surprising that of the six ruptures, which from the formal point of view are analogous, only one is perceived by the scientific community as a revolution while the remaining five are seen as organic growth. The question is then: *why is the discovery of non-Euclidean geometry interpreted so differently from other, formally analogous, ruptures?*

It seems that whether a rupture is classified as a revolution or not does not depend solely on intra-scientific reasons. The discovery of topology was as important as the discovery of non-Euclidean geometry. Thus there can be no internal reason to classify one change as revolutionary and the other as not. The discovery of non-Euclidean geometry obtained its high social charge at least to a degree from the role it played in the refutation of Kantian philosophy. If it is really so, then a better understanding of the question of why a particular epistemic rupture is viewed as a revolution could shed new light on the ways in which science functions in culture, what illusions and expectations society associates with science and when it is forced to give up these illusions. So separation of the epistemological and the sociological aspects of the development of science and their subsequent confrontation opens new insights. Separation of the epistemological and the sociological aspects of the changes in science resembles the separation of the context of discovery and the context of justification in the first half of the twentieth century. Nevertheless, when the neo-positivists introduced this distinction, since they never tried to relate these two contexts, they deprived themselves of the most interesting insights which such a distinction can

bring. Often only after a separation of the epistemological and the sociological (or of the logical and psychological) does it become possible
to formulate the most important questions.

4.1.3. Possible Refinements of the Kuhnian Picture

The main point that can be raised against Kuhn's theory on the basis of our analyses is that his notion of scientific revolutions includes
changes of different kinds.[1]   Kuhn includes among revolutions *re-
codings*, which he analyzes on the example of the Copernican revolution, as well as *relativizations*, an example of which is the oxida-
tion theory of combustion. Therefore his categories, such as paradigm,
anomaly, crisis, and revolution that he obtained from the analysis of
such heterogeneous material, are rather rough and unspecific. Each
of Kuhn's categories includes several different notions. It seems that
something different comprises the paradigm in the case of a re-coding
from that in the case of a relativization. Due to this unspecific nature
of Kuhn's analyses, the picture of scientific revolutions that he offers
does not capture the epistemological structure of scientific revolutions
but is rather a description of the process of adaptation of the scientific
community to change. He chose the *scientific* community, but similar processes as those described by Kuhn can be found also in other
communities, when they are confronted with radical failure of their
previous adaptation strategies. It is likely that if a *peasant* community would be confronted with a series of bad harvests (*anomalies* in
contrast to *normal farming*) a *crisis* could result. Some farmers would
try to grow different crops than they used to grow (they would deviate from the *paradigm*). If some of the new crops proved successful,
it would spread throughout the *peasant community* (many adherents of
the traditional crop would *convert* to the new one) and *normal farming*,
based on a new crop (the *new paradigm*) would be reestablished. Thus
it seems that Kuhn did not describe the structure of *scientific* revolu-

---

[1] The discrimination of re-codings and relativizations which were introduced in the reconstruction of
the development of mathematics can be introduced also to the field of physics. Thus the births of field
theory or of quantum mechanics were re-codings while the births of the Lagrangean or Hamiltonian
mechanics were two relativizations in classical mechanics.

tions, but he has only illustrated with the example of science a general strategy of social adaptation.[2]

In order to be able to characterize more precisely the basic categories of Kuhn's theory and to grasp the specific features that distinguish scientific revolutions from other processes of social adaptation, it seems to be necessary to distinguish different kinds of scientific revolutions. I suggest distinguishing the *paradigm of coding*, which specifies the particular code by means of which science represents reality, from the *paradigm of positing*, which specifies what from all that the code allows to be represented, gets the status of being absolute (i.e., given, fundamental, unquestionable) and what is relative (i.e., constructed, derived, questionable). It is probable that two paradigms of coding have different degrees of incommensurability from two paradigms of positing. Maybe in this way we can understand the phenomenon of incommensurability more clearly. A particular kind of scientific revolution would then consist in changes of the paradigms of that particular kind. A re-coding, viewed this time not as an epistemic rupture but as a scientific revolution accompanying this rupture, would be a change of the paradigm of coding, similarly as a relativization would be a change of the paradigm of positing.

Nevertheless, in the course of a re-coding, not only the paradigm of coding is being changed, but radical changes occur also in the paradigm of positing. This is so because a re-coding is such a deep and radical change that it destabilizes also the positing. This fact makes it possible to describe in more detail the epistemological dynamics of the scientific revolution of a particular kind. A revolution is not a singular act, but it consists rather of a process of gradual changes, which besides the main rupture that determines the character of the revolution contains also several ruptures of smaller magnitudes. In each particular kind of revolution the anomalies are of a different nature (in the case of re-coding the anomalies have the form of the logical and expressive boundaries) and the crisis has a different depth. The analysis of all these aspects makes it possible to describe what Giorello has called *the fine structure of scientific revolutions* (Giorello 1992). It seems probable that the fine structure of a re-coding will consist of a series of relativizations that leads to it, similarly as the fine structure of a rela-

---

[2] Of course, the generality of Kuhn's theory might be seen as a point in its favor rather than against. After all, once the general picture has been sketched, details can be added. Thus our criticism of Kuhn can be seen more appropriately as suggestions of some refinements of the overall picture he presented.

tivization will consist of a series of re-formulations. The hierarchical order of the paradigms of different kinds and the fine structure of scientific revolutions that result from this hierarchical order are phenomena that have no analogy in other processes of social adaptation. This hierarchical structure is peculiar to scientific revolutions and distinguishes them from social changes, say in the area of religion. The fact that the scientific revolutions were often compared to religious conversions was only a consequence of the fact that Kuhn grounded the dynamics of the scientific revolution in the scientific community, i.e., in its sociological aspect. From the sociological point of view a scientific community reacts to changes in a similar way as a religious community. But if we turn to the fine structure of scientific revolutions, this will distinguish a scientific revolution from any religious conversion.

4.2. Mathematical Research Programmes (Lakatos)

We can use our reconstruction of the development of synthetic geometry to explain a strange aspect of Lakatos' *Proofs and Refutations* (Lakatos 1976). In his book Lakatos offered a rational reconstruction of the historical development of the theory of polyhedra. Besides the appendixes the book consists of two major parts. In the first, geometrical, part Lakatos presents a detailed analysis of *monster-barring, exception-barring*, and *lemma-incorporation*. There is no doubt that mathematicians of the past did use the mentioned strategies and thus their philosophical reconstruction is an important contribution to our understanding of how mathematics really works. The second major part of the book, containing Poincaré's topological proof of Euler's theorem, is not so thoroughly worked out. Nevertheless we can assume that if Lakatos had lived longer, he would have displayed his gamut of heuristic strategies of *monster-barring, exception-barring*, and *lemma-incorporation* also on this material.

For similar reasons we can omit a mistake on page 123 of *Proofs and Refutations*, where Lakatos claims that the heptahedron (i.e., the projective plane) bounds. Here, Lakatos contradicts his own statement from page 120 of the same book, where he (correctly) asserted, that the heptahedron is a one-sided surface for which reason there exists no geometrical body, which the heptahedron could bound. The gaps in Lakatos' original presentation were filled in by Mark Steiner in his paper *The philosophy of mathematics of Imre Lakatos* (Steiner 1983) where he presented an exposition of Poincaré's proof. The inconsistency in the second part of the *Proofs and Refutations* is probably a result of the fact that Lakatos did not want to complicate his text with details concerning the orientation of surfaces. But if we define the concept of boundary without orientation, the whole theory becomes obscure, and it is easy to make mistakes. Steiner introduced the concept of orientation, at least implicitly, when speaking of the "*sum*" and "*difference*" of schemes (Steiner 1983, p. 515).

However this is not the problem that concerns us here. What is striking is the lack of any attempt to connect the two parts of the book. Lakatos, who always stressed the necessity of reconstructing the circumstances in which the new concepts emerged, and criticizes mathematicians – like Hilbert or Rudin – who presented formal definitions without any historical background (see Lakatos 1976, pp. 15 and 145), suddenly pulls out of the "top-hat" the basic concepts of algebraic topology without the slightest comment, and pretends that everything is all right.

4.2.1. The Lack of Connection between the Two Major Parts of Proofs and Refutations

Our reconstruction of the development of synthetic geometry enables us to explain this aspect of Lakatos' book. The transition from geometry to algebraic topology is a relativization, consisting in the introduction of the constitutive form of language. This transition, nevertheless, did not occur primarily in the theory of polyhedra (i.e., in the theory, the development of which Lakatos analyzed), but in the theory of functions of complex variables. Poincaré was led to the fundamental changes of language by problems in complex analysis. He then transferred the conceptual progress, achieved in complex analysis into geometry, and demonstrated the power of the new language in the theory of polyhedra.

A relativization – one which had its origin outside the theory which Lakatos was studying – was a rupture, with which his methods of analysis did not enable him to deal. His interpretative tools such as *monster-barring* or *lemma-incorporation* failed, because here we are dealing with a change of the whole conceptual basis of geometry and not just with the assumptions of some theorem. Thus, what appeared to be only a mere omission sheds light onto the boundaries of applicability of Lakatos' method of reconstruction. *Lakatos' method can only be adopted in cases where the form of language is not changing.* So even if, in the last chapters of his book, Lakatos is writing about concept formation, the fact that he did not recognize one of the most fundamental conceptual changes in the history of geometry casts doubts concerning the reliability of this part of his analysis.

After what we have said, it is not surprising that Lakatos missed the epistemological nature of Poincaré's proof. He presents it as a *"translation of the conjecture into the terms of vector-algebra"* (p. 106). But the following questions remained unanswered: Why into the terms of *vector-algebra*? Where is it coming from? And why into vector-algebra *modulo 2*? Why not modulo 3 or 7? From the reconstruction of Klein's Erlanger program we know that Poincaré, just like Klein, made no translation. The birth of combinatorial topology was something more radical than a translation. Poincaré made explicit the constitutive acts, forming the basis of the concept of a topological space. The fact that these acts are based on arithmetic modulo 2 shows that we have to deal here with constitutive acts. Arithmetic modulo 2 expresses parity. Parity means that the chain of acts "closes itself", and so constitutes an object (an open chain does not constitute anything).

4.2.2. Reduction of the Development of Mathematics to Re-Formulations

Later Lakatos turned from the methodology of proofs and refutations to the concept of scientific research programmes. The notion of the "research programme" appeared for the first time in the paper *Changes in the Problems of Inductive Logic* (Lakatos 1968a). Lakatos used this notion in order to discuss Carnap's research programme in epistemology: "A successful research programme bustles with activity. There are dozens of puzzles to be solved and technical questions to be answered; even if some of these - inevitably - are the programme's own creation." He presented the methodology of scientific research programmes in his paper *Criticism and the Methodology of Scientific Research Programmes* (Lakatos 1968b). According to it, great scientific achievements are not characterised by an *isolated hypothesis* (here Lakatos is moving away from the position of *Proofs and Refutations*, where he presented the development of the theory of polyhedra as centred around Euler's formula), but rather by a *research programme*. A research programme manifests itself in the form of a series of successively developed scientific theories, characterised by a certain continuity which connects the members of the series.

But even after turning to the methodology of scientific research programmes, he still *remained on the level of re-formulations.* In his *Falsification and the Methodology of Scientific Research Programmes* Lakatos gives a detailed description of Newton's programme:

> "Newton first worked out his programme for a planetary system with a fixed point-like sun and one single point-like planet. It was in this model that he derived his inverse square law for Kepler's ellipse. But this model was forbidden by Newton's own third law of dynamics, therefore the model had to be replaced by one in which both sun and planet revolved round their common centre of gravity … Then he worked out the programme for more planets as if there were only heliocentric but no interplanetary forces. Then he worked out the case where the sun and planets were not mass-points but mass-balls … This change involved considerable mathematical difficulties, held up Newton's work – and delayed the publication of the Principia by more than a decade. Having solved this 'puzzle', he started work on spinning balls and their wobbles. Then he admitted interplanetary forces and started work on perturbations."
> (Lakatos 1970, p. 50)

In contrast to this detailed description of Newton's programme, Lakatos says nothing about passing from Newtonian mechanics to Lagrangean, and later to Hamiltonian. So even if he mentioned the possibility of programmes on larger scales, he never gave any example of such a programme. This seems strange, if we have in mind that Lakatos conceived his theory as an alternative to Kuhn's theory of scientific revolutions, which operates on the large scale of science as a whole.

The reason for the omission of the Lagrangean and Hamiltonian mechanics is similar to the reason of the gap between the geometrical and topological parts of *Proofs and Refutations*. The transitions to Lagrangean and to Hamiltonian mechanics are relativizations, and they cannot be reconstructed by the means used by Lakatos. Thus even if Lakatos claims that his aim is a rational reconstruction of the development of science, he ignored all processes in the course of which more significant changes would occur. He did not analyse any re-coding; and for relativizations he either ignored them (as in mechanics) or misinterpreted them (as in the case of topology). His attention was completely absorbed by the details of re-formulations and other kinds of change escaped him.

4.3. Stages of Cognitive Development (Piaget)

Piaget and Garcia offered in their book *Psychogenesis and the History of Science* (Piaget and Garcia 1989) an epistemological reconstruction of the development of geometry, algebra, and classical mechanics. In the history of geometry they singled out three periods: Euclidean geometry, projective geometry, and Klein's Erlanger program. They associated these periods with the three developmental stages of geometrical thought: *intra-figurative*, *inter-figurative*, and *trans-figurative*. The book contains a detailed analysis of processes of transformation of knowledge which mediate the transition from one stage in the development of geometry to the next one. The epistemological analyses of Piaget and Garcia are underpinned with psychological research that show that analogous stages to those which the authors discriminated in the history of geometry can be found also in the cognitive development of children. But despite these compelling aspects the epistemological reconstructions of Piaget and Garcia give rise also to some concern.

4.3.1. The Problem of the Omitted Stages

When we confront the theory of the development of geometry, presented by Piaget and Garcia, with the actual history of geometry, some doubts appear. First of all, the authors *ignore certain stages of the development of geometry* (for instance non-Euclidean geometry) which from the historical point of view were by no means less important than the stages which they adduce (projective geometry). Klein's Erlanger program, by which Piaget and Garcia close their reconstruction of the development of geometry, was a reaction to the discovery of the non-Euclidean geometries. Therefore by ignoring it they cannot reconstruct the motives for what Klein was doing. Secondly they *present the development of geometry as closed*. After reaching the trans-figurative stage there is no further stage that could follow. Our reconstruction of the development of geometry makes it possible to overcome both of these limitations. In its framework, non-Euclidean geometry gets its proper place in the development of geometry that is on a par with the other stages and after the integrative form (that corresponds to Klein's Erlanger program). The development continues further to the constitutive and conceptual stages. This shows that these two aspects of the theory of Piaget and Garcia do not have footing in the history of geometry. The historical development of geometry gives us no right to ignore

Lobachevski or to pronounce Klein to be the final stage of the development of geometry. Lobachevski belongs to the history of geometry with the same right as Poncelet or Klein. Thus the question arises, why did the authors omit Lobachevski from the development of geometry? The answer is not difficult to find. Their three stages (*intra-, inter-,* and *trans-*) resemble the Hegelian triad (*thesis, antithesis, synthesis*). Thus the inaccuracies of Piaget's and Garcia's analysis of the development of geometry have their origin not in the historical material but in the conceptual scheme by means of which they analyze it.

4.3.2. The Relation of Ontogenesis and Phylogenesis

Another controversial aspect of the analyses of the development of geometry presented by Piaget and Garcia is their thesis that the ontogenesis recapitulates *the historical development in reversed order* (Piaget and Garcia 1989, p. 113). According to this thesis, children first master the topological invariants and only afterwards do they master the metric ones, while historically the metric (Euclidean) geometry was formulated first and only much later did topology appear. The reliability of this claim is weakened by the fact that in their reconstruction of the development of geometry Piaget and Garcia did not mention topology. Thus they are attempting to establish an analogy between the results of psychological research and the historical material that they did not analyze. Of course, at first sight the drawings of small children *appear* analogous to topology. But they only appear as such. If we keep in mind the complex epistemological structure of the language of topology, it is rather improbable that the children could really master it before the Euclidean stage.

If we examine the structure of Piaget's and Garcia's book more closely, we can find an interesting peculiarity. The authors reconstruct the development of a scientific discipline either only before it became an exact science (thus in the case of mechanics they describe only the pre-Newtonian period), or they only reconstruct the period of the discipline as an exact science (the chapters on geometry and on algebra). Nevertheless, none of their reconstructions includes the process of constitution of exactness, i.e., of the process that would include both the pre-paradigmatic as well as the paradigmatic stages of development of the particular discipline. Thus in the case of mechanics they do not connect the pre-Newtonian period (including Aristotle, Oresme, and Buridan) with the post-Newtonian period (including Euler, Lagrange, and

Hamilton). In the case of mechanics they confine their analysis to the first of these two periods. Similarly in the reconstruction of the development of geometry they do not connect the pre-Euclidean period (including Egyptian and Babylonian geometry) with the post-Euclidean period (including Desargues, Poncelet, and Klein). Therefore it seems likely that *proto-geometry* (what we could perhaps call the geometrical thought of children before they master metric geometry) has its historical analogy rather in the geometry of the ancient Egyptians than in topology. Piaget and Garcia are forced to relate proto-geometry to topology because they excluded the pre-Euclidean period from their reconstruction of the history of geometry. Then, from all that remained, topology seemed to be most similar to the geometrical thought of children.

But let us come back to the thesis about the reversed order of the ontogenesis of geometry. We have seen that it is not a "deep" analogy, i.e., an analogy that would be based on the similarity of the epistemological structure of topology and the geometrical thought of small children. On the other hand we cannot deny that there is some similarity between the drawings of children and the figures that can be found in books on topology. It is not a similarity in a strong sense. It does not mean that when a child draws a house, it preserves all the topological invariants (the number of components, orientation, and continuity) but violates the metric invariants (proportions, angles). Usually children violate also the topological invariants. Many lines that should be continuous they draw as divided; several areas that should be closed they draw open. Thus the analogy does not work in detail. But it cannot be denied there is some similarity. The basis of this similarity is the fact that the topological invariants are much more robust than the metric ones. If we deform a circle into an ellipse, from the topological and the projective point of view nothing happened, while metrically seen the object was changed. Similarly if we deform a circle into a square, from the point of view of topology nothing happened, while from the projective and metric point of view it is a radical change. Therefore it is natural to assume a stronger degree of conservation of the topological invariants than of the projective or metric ones in every learning process. Always when a child has to learn some complex coordination it masters first its robust topological structure, then its projective structure (the keeping of parallels), and only at the end is the child able to master also its subtle metric structure.

Bibliography

M. Agoston, (1976): *Algebraic Topology, a First Course*, Marcel Dekker, New York.

M. Aigner and G. M. Ziegler, (1998): *Proofs from The Book*, Springer, Berlin.

L. B. Alberti, (1435): *De pictura*, English translation by J. R. Spencer: *On Painting*, Yale University Press, New Haven 1956.

S. Albeviero, J. E. Fenstad, R. Hoegh-Krohn and T. Lindstrom, (1986): *Nonstandard Methods in Stochastic Analysis and Mathematical Physics*. Academic Press, London.

M. I. M. Al-Khwárizmí, (850?): *Matematicheskije traktaty*, (Mathematical treatises, in Russian.) FAN, Tashkent 1983.

H.-W. Alten, A. Djafari Naini, M. Folkerts, H. Schlosser, K.-H. Scholte and H. Wussing, (2003): *4000 Jahre Algebra, Geschichte, Kulturen, Menschen*, Springer, Berlin 2005.

Archimedes: The Works of Archimedes. in: *Great Books of the Western World*, vol. 11, *Encyclopedia Britannica*, Chicago 1952, pp. 403–592.

L. Arkeryd, N. Cutland and C. W. Henson, (1997): *Nonstandard Analysis, Theory and Applications*, Kluwer Academic Publishers, Dordrecht.

J. Barrow-Green, (1997): *Poincaré and the Three Body Problem*, American Mathematical Society.

E. Beltrami, (1868): *Saggio di interpretazione della geometria noneuclidea*, Giorn. Mat. Battaglini, **6**, pp. 284–312. Russian translation in: Norden (1956), pp. 180–212.

G. Berkeley, (1734): *The Analyst; or, a Discourse adressed to an Infidel Mathematician* ... in: A. A. Luce, T. E. Jessup (eds.), *The Works of George Berkeley, Bishop of Cloyne*, **4**, London 1951.

P. Bernays, (1926): Axiomatische Untersuchung des Aussagen-Kalkuls der "Principia mathematica", *Mathematische Zeitschrift*, **25**, pp. 305–320.

L. Boi, (1995): *Le Probléme mathématique de l'espace*, Springer, Berlin.

L. Boi, D. Flament and J.-M. Salanskis, eds. (1992): *1830–1930 A Century of Geometry*, Springer, Berlin.

J. Bolyai, (1831): *Appendix. The Science of Absolute Space*, English translation by G. B. Halsted, published as a supplement in: Bonola (1906).

B. Bolzano, (1817): *Rein analytischer Beweis des Lehrsatzes, dass zwischen je zwei Werten, die ein entgegengesetztes Resultat gewähren, wenigstens eine reelle Wurzel der Gleichung liege*, English translation by S. B. Russ, *Historia Mathematica*, **7** (1980), pp. 156–185.

R. Bonola, (1906): *Non-Euclidean Geometry, with a supplement containing "The theory of Parallels" by N. Lobachevski and "The science of absolute space" by J. Bolyai.* Dover, New York 1955.

G. Boole, (1847): *The Mathematical Analysis of Logic*, Cambridge.

G. S. Boolos and R. C. Jeffrey, (1974): *Computability and Logic*, Cambridge University Press 1994.

C. Boyer and U. Merzbach (1989): *A history of Mathematics*, John Wiley, New York.

G. Cantor, (1872): Über die Ausdehnung eines Satzes aus der Theorie der trigonometrischen Reihen, *Mathematische Annalen*, **5**, 123–132.

G. Cantor, (1880): Über unendliche lineare Punktmannigfaltigkeiten. No. 2. *Math. Ann.*, **17**, 355–358.

G. Cantor, (1883): *Grundlagen einer allgemeiner Mannigfaltigkeitslehre - Ein mathematisch-philo-sophischer Versuch in der Lehre des Unendlichen*, Teubner, Leipzig. English translation in: Ewald (1996), pp. 878–920.

G. Cantor, (1932): *Gesammelte Abhandlungen mathematischen und philosophischen Inhalts*, E. Zermelo, ed., J. Springer, Berlin.

G. Cardano, (1545): *Ars Magna, or the Rules of Algebra*, MIT Press 1968.

L. Carnot, (1797): *Réflexions sur la métaphysique du calcul infinitésimal.* Paris.

A. L. Cauchy, (1821): *Cours d'Analyse de l'École Polytechnique. Premiere partie. Analyse algébrique*, Paris.

A. Cayley, (1859): *A sixth memoir upon quantics*, Philosophical Transactions of the Royal Society of London, vol. 149, pp. 61–90. Russian translation in: Norden (1956), pp. 222–252.

A. Church, (1969): Review of: The Completeness of Elementary Algebra and Geometry. *The Journal of Symbolic Logic*, **34**, pp. 302.

D. Corfield, (1997): Assaying Lakatos's Philosophy of Mathematics, *Studies in the History and Philosophy of Science*, **28**, pp. 99–121.

D. Corfield, (2003): *Towards a Philosophy of Real Mathematics*, Cambridge University Press, Cambridge.

R. Courant, (1927): *Differential and Integral Calculus I*, Wiley 1988.

R. Courant and H. Robbins, (1941): *What is mathematics?* Oxford University Press, New York 1978.

M. Crowe, (1975): Ten "laws" concerning patterns of change in the history of mathematics. *Historia Mathematica*, **2**, pp. 161–166. Reprinted in: Gillies 1992, pp. 15–20.

N. Cusanus, (1440): De Docta Ignorantia. English translation by J. Hopkins in: *Complete Philosophical and Theological Treatises of Nicholas of Cusa*, vol. 1. The Arthur J. Banning Press, Minneapolis 2001.

J. Dauben, (1979): *Georg Cantor. His Mathematics and Philosophy of the Infinite*, Princeton University Press, Princeton.

J. Dauben, (1984): Conceptual revolutions and the history of mathematics: two studies in the growth of knowledge. in: *Transformation and Tradition in the Science* E. Mendelsohn, ed., pp. 81–103, Cambridge University Press. Reprinted in: Gillies 1992, pp. 49–71.

P. J. Davis and R. Hersh, (1983): *The Mathematical Experience*, Birkhäuser, Boston.

R. Dedekind, (1872): *Stetigkeit und Irrationale Zahlen.* Friedrich Vieweg, Braunschweig 1965. English translation: *Continuity and Irrational Numbers*, in: Dedekind (1901), pp. 1–27.

R. Dedekind, (1888): *Was sind und was sollen die Zahlen?* Friedrich Vieweg, Braunschweig 1965. English translation: *The Nature and Meaning of Numbers*, in: Dedekind (1901), pp. 31–115.

R. Dedekind, (1901): *Essays on the Theory of Numbers*, English translation by W. W. Beman, The Open Court Publishing Company. Reprinted by Dover, New York 1963.

R. Descartes, (1637): *The Geometry of Rene Descartes*, Dover, New York 1954.

F. Diacu and P. Holmes, (1996): *Celestial Encounters. The Origins of Chaos and Stability*, Princeton University Press, Princeton.

H. Dörrie, (1932): *Triumph der Mathematik.* Physica Verlag, Würzburg 1958.

M. Dummett, (1994): *Origins of Analytic Philosophy*, Harvard University Press, Cambridge, Mass.

C. Dunmore, (1992): Meta-level revolutions in mathematics. in: Gillies 1992, pp. 209–225.

C. H. Edwards, (1979): *The Historical Development of the Calculus*, Springer, New York.

Euclid: *The Thirteen Books of the Elements*, Translated by Sir Thomas Heath, Dover, New York 1956.

L. Euler, (1748): *Introductio in analysin infinitorum*, Bousquet, Lausannae. English translation by J. Blanton: *Introduction to Analysis of the Infinite*, Springer Verlag 1988.

L. Euler, (1770): *Vollständige Anleitung zur Algebra*, Reclam, Leipzig 1911. English translation by J. Hewlett: *Elements of Algebra*, Longman, London 1822.

W. Ewald, (1996): *From Kant to Hilbert: A Source Book in the Foundations of Mathematics*, Clarendon Press, Oxford.

J. Fauvel, ed. (1993): *Newtons Werk. Die Begründung der modernen Naturwissenschaft*, Birkhäuser, Basel.

J. Fauvel and J. Gray, (1987): *The History of Mathematics: A Reader*, Macmillan, London.

J. Ferreirós, (1999): *Labyrinth of Thought. A History of Set Theory and its Role in Modern Mathematics*, Birkhäuser, Basel.

J. Fourier, (1822): *The Analytic Theory of Heat*, English translation by A. Freeman, Dover, New York 2003.

A. Fraenkel, (1922): Der Begriff "definit" und die Unabhängigkeit des Auswahlsaxioms. *Sitzungsberichte der Preussischen Akademie der Wissenschaften*, pp. 253–257. English translation in: van Heijenoort 1967, pp. 284–289.

A. Fraenkel, (1928): *Einleitung in die Mengenlehre*, Springer, Berlin.

A. Fraenkel and Y. Bar-Hillel, (1958): *Foundations of Set Theory*, North-Holland, Amsterdam.

G. Frege, (1879): *Begriffsschrift, eine der arithmetischen nachgebildete Formelsprache des reinen Denkens*, Georg Olms, Hildesheim 1993. English translation in: van Heijenoort 1967, pp. 1–82.

G. Frege, (1884): *Die Grundlagen der Arithmetik.* Breslau. English translation by J. L. Austin, *The Foundations of Arithmetic*, Blackwell, Oxford 1953.

G. Frege, (1891): Funktion und Begriff. in: G. Frege,: *Funktion, Begriff, Bedeutung*, Vandenhoek & Ruprecht, Göttingen 1989, pp. 17–39. English translation in: P. Geach and M. Black, (eds.): *Translations from the Philosophical Writings of Gottlob Frege*, pp. 21–41.

G. Frege, (1893): *Grundgesetze der Arithmetik, Begriffsschriftlich abgeleitet*, Olms, Hildesheim 1962.

M. Friedman, (1985): Kant's theory of geometry. *The Philosophical Review*, **94**, pp. 456–506.

M. Friedman, (1992): *Kant and the Exact Sciences*, Harvard University Press, Cambridge.

A. D. Gelfond, (1952): *Transcendental and Algebraic Numbers*, Translated by L. Boron, Dover, New York 1960.

D. Gillies, (1982): *Frege, Dedekind and Peano on the Foundations of Arithmetic*, Van Gorcum, Assen.

D. Gillies, ed. (1992): *Revolutions in Mathematics*, Clarendon Press, Oxford.

D. Gillies, (1992a): *The Fregean revolution in Logic*, in: Gillies 1992, pp. 265–306.

G. Giorello, (1992): The "fine structure" of mathematical revolutions: metaphysics, legitimacy, and rigor. *The case of the calculus from Newton to Berkeley and Maclaurin*, in: Gillies 1992, pp. 134–168.

K. Gödel, (1930): Die Vollständigkeit der Axiome des logischen Funktionenkalküls. *Monatshefte für Mathematik und Physik* **37**, 349–360. English translation in: van Heijenoort 1967, pp. 582–591.

K. Gödel, (1931): Über formal unentscheidbare Sätze der Principia mathematica und verwandter Systeme I. *Monatshefte für Mathematik und Physik* **38**, 173–198. English translation in: van Heijenoort 1967, pp. 596–616.

I. Grattan-Guiness, ed. (1992): *Companion Encyclopedia of the History and Philosophy of the Mathematical Sciences*, Routledge, London.

J. Gray, (1979): *Ideas of Space Euclidean, Non-Euclidean, and Relativistic*, Clarendon Press, Oxford.

J. Gray, (1992): The nineteenth-century revolution in mathematical ontology. in: Gillies 1992, pp. 226–248.

M. T. Graves, (1997): *The Philosophical Status of Diagrams*, PhD Thesis. Stanford University.

E. Grosholz, (1992): *Was Leibniz a mathematical revolutionary?* in: Gillies 1992, pp. 117–133.

E. Grosholz and H. Bregger, (eds. 2000): *The Growth of Mathematical Knowledge*, Kluwer.

J. Hadamard, (1945): *An essay on the Psychology of Invention in the Mathematical Field*, Princeton University Press.

G. H. Hardy and W. W. Rogosinski, (1944): *Fourier Series*, Dover, New York 1999.

F. Hausdorff, (1914): *Grundzüge der Mengenlehre*, Leipzig.

T. Heath, (1921): *A History of Greek Mathematics*, Dover, New York 1981.

D. Hilbert, (1899): *Grundlagen der Geometrie*, Teubner, Leipzig 1930. English translation by E. J. Townsend: *The Foundations of Geometry*, The Open Court Publishing Company, La Salle 1950.

J. Hintikka, (1965): Kant's new method of thought and his theory of mathematics. *Ajatus*, **27**, 37–43. Reprinted in: *Knowledge and the Known, Modern Essays*, Reidel, pp. 126–134.

J. Hintikka, (1967): Kant on the Mathematical Method, *The Monist*, **51**, pp. 352–375. Reprinted in: C. J. Posy, ed. (1992), pp. 21–42.

G. Ifrah, (1981): *Universalgeschichte der Zahlen*, Campus, Frankfurt 1986.

H. N. Jahnke, ed. (1999): *Geschichte der Analysis*, Spektrum, Heidelberg.

F. Kaderavek, (1922): *Perspektiva. Příručka pro architekty, malíře a přátele umění*, (Perspective, a Handbook for Architects, Painters, and Friends of Art, in Czech. Jan Štenc, Praha.

V. Kagan, (1949): *Osnovanija geometrii*, (Foundations of Geometry, in Russian.) GITTL, Leningrad.

J. M. Keynes, (1947): Newton, the man. in: *The Royal Society Newton Tencentenary Celebrations*, Cambridge University Press 1947, pp. 27–34.

F. Klein, (1872): *Vergleichende Betrachtungen über neuere geometrische Forschungen (Das Erlanger Program)*. Deichart.

F. Klein, (1928): *Vorlesungen über nicht-euklidische geometrie*, Springer, Berlin.

J. Klein, (1934): *Greek Mathematical Thought and the Origin of Algebra*, MIT Press 1968.

M. Kline, (1972): *Mathematical Thought from Ancient to Modern Times*, Oxford University Press, New York.

W. Kneale and M. Kneale, (1962): *The Development of Logic*, Oxford University Press.

E. Knobloch, (2002): Unendlichkeit und Mathematik bei Nicolaus von Kues. in: Schürmann, A. and B. Weiss, (eds.): *Chemie-Kultur-Geschichte, Festschrift für Hans-Werner Schütt*, Verlag für Geschichte der Naturwissenschaften und der Technik, Berlin 2002, pp. 223–234.

W. Knorr, (1986): *The Ancient Tradition of Mathematical Problems*, Birkhäuser, Boston.

T. Koetsier, (1991): *Lakatos' Philosophy of Mathematics, A Historical Approach*, North-Holland Amsterdam.

A. Kolman, (1961): *Dejiny matematiky ve staroveku*, (History of Mathematics in Antiquity.) Academie, Praha 1968.

T. S. Kuhn, (1962): *The Structure of Scientific Revolutions*, University of Chicago press.

T. S. Kuhn, (1974): Second Thoughts on Paradigms. in: *The Essential Tension. Selected studies in scientific tradition and change*, The University of Chicago Press 1977, pp. 293–319.

L. Kvasz, (1998): History of Geometry and the Development of the Form of its Language. *Synthese*, **116**, 141–186.

L. Kvasz, (1999): On classification of scientific revolutions. *Journal for General Philosophy of Science*, **30**, 201–232.

L. Kvasz, (2000): Changes of Language in the Development of Mathematics. *Philosophia mathematica*, **8**, 47–83.

L. Kvasz, (2002): Lakatos' Methodology Between Logic and Dialectic. in: *Appraising Lakatos*, Eds. G. Kampis, L. Kvasz and M. Stöltzner, Kluwer, pp. 211–241.

L. Kvasz, (2004): The Invisible Dialog Between Mathematics and Theology. in: *Perspectives on Science and Christian Faith*, **56**, pp. 111–116.

L. Kvasz, (2005): Similarities and differences between the development of geometry and of algebra. in: *Mathematical Reasoning and Heuristics*, (C. Cellucci and D. Gillies, eds.), King's College Publications, London 2005, pp. 25–47.

L. Kvasz, (2006): History of Algebra and the Development of the Form of its Language. *Philosophia Mathematica*, **14**, pp. 287–317.

L. Kvasz, (2007): Visual Illusions in painting, or What could Computer Graphics Learn from Art History. *Proceedings of the Spring Conference on Computer Graphics SCCG 2007*, ACM Press, pp. 17–30.

J. L. Lagrange, (1788): *Mécanique Analytique*, Paris. English translation by A. Boissonnade and V. N. Vagliente: *Analytical Mechanics*, Boston Sudies in the Philosophy of Science, Kluwer 1998.

J. L. Lagrange, (1797): *Théorie des fonctions analytiques*, Paris.

I. Lakatos, (1963–64): Proofs and Refutations. *British Journal for the Philosophy of Science*, **14**, pp. 1–25, 120–139, 221–243, 296, 342.

I. Lakatos, (1968a): Changes in the Problems of Inductive Logic. in: *Mathematics, Science and Epistemology. Philosophical Papers of Imre Lakatos*, vol. 2, Cambridge University Press, Cambridge 1978, pp. 128–200.

I. Lakatos, (1968b): Criticism and the Methodology of Scientific Research Programmes. in: *Proceedings of the Aristotelian Society*, **69**, pp. 149–186.

I. Lakatos, (1970): Falsification and the methodology of scientific research programmes. in: *The methodology of scientific research programmes. Philosophical Papers of Imre Lakatos*, vol. I. Cambridge University Press, Cambridge 1978, pp. 8–101.

I. Lakatos, (1976): *Proofs and Refutations*. Cambridge University Press, Cambridge.

G. W. Leibniz, (1686): *Discourse on Metaphysics*, Translated by P. G. Lucas and L. Grint. Manchester University Press, Manchester 1953.

N. I. Lobachevski, (1829): O nacalach geometrii. (On the Foundations of Geometry, in Russian.) in: *Polnoje sobranie socinenij*, tom 1 (*Collected Works*, vol. 1), GITTL, Leningrad 1946.

P. Mancosu, (1992): Descartes's Géometrie and revolutions in mathematics. in: Gillies 1992, pp. 83–116.

B. Mandelbrot, (1967): How long is the coast of Britain? Statistical self-similarity and fractional dimension. *Science*, **155**, 636–638.

B. Mandelbrot, (1977): *The Fractal Geometry of Nature*, Freeman, New York 1982.

J. H. Manheim, (1964): *The Genesis of Point Set Topology*, Pergamon Press, Oxford.

F. A. Medvedev, (1965): *Razvitije teorii mnozhestv v XIX veke*, (Development of set Theory in the XIX Century, in Russian). Nauka, Moskva.

H. Mehrtens, (1976): T. S. Kuhn's theories and mathematics: a discussion paper on the 'new historiography' of mathematics. *Historia Mathematica*, **3**, 297–320. Reprinted in: Gillies 1992, pp. 21–41.

A. F. Monna, (1975): *Dirichlet's principle a mathematical comedy of errors and its influence on the development of analysis*, Oosthoek, Scheltema and Holkema, Utrecht.

G. H. Moore, (1982): Zermelo's Axiom of Choice, its Origins, Development, and Influence. Springer, New York.

I. Newton, (1687): *The Principia*, A New Translation by I. B. Cohen and A. Whitman, Proceeded by *A Guide to Newton's Principia*, University of California Press, Berkeley 1999.

A. P. Norden, ed. (1956): *Ob osnovanijach geometrii.* (on the Foundations of Geometry, in Russian.) GITTL, Moskva.

G. Olivieri, (2005): Do We Really Need Axioms in Mathematics? in: *Mathematical Reasoning and Heuristics*, (C. Cellucci and D. Gillies, eds.), King's College Publications, London, pp. 119–135.

M. Pasch, (1882): *Vorlesungen über neuere Geometrie*, Teubner, Leipzig.

G. Peano, (1889): *Arithmetices principia nova methodo exposita*, English translation in: H. C. Kennedy, ed., *Selected Works of Giuseppe Peano*, Allen and Unwin 1973.

H. O. Peitgen and P. H. Richter, (1986): *The Beauty of Fractals*, Springer, Heidelberg.

H. O. Peitgen, H. Jürgens and J. Saupe, (1992): *Chaos and Fractals*, Springer, New York.

J. Piaget and R. Garcia, (1983): *Psychogenesis and the history of science*, Columbia University Press, New York 1989.

H. Poincaré, (1890): Sur le probléme des trois corps et les équations de la dynamique, *Acta Mathematica*, **13**, 1–270.

H. Poincaré, (1895): Analysis situs. *J. École Polytechniques*, Cahier 1, pp. 1–121.

H. Poincaré, (1899): Complément á l'Analysis situs. *Rendiconti Circolo mat. Palermo*, **13**, pp. 285–343.

H. Poincaré, (1902): *La Science et l'Hypothése*, Flammarion, Paris. English translation in: *The Value of Science: Essential Writings of Henri Poincaré*, Random House, New York 2001.

H. Poincaré, (1908): *Science et Méthode*, Flammarion, Paris. English translation in: *The Value of Science: Essential Writings of Henri Poincaré*, Random House, New York 2001.

G. Polya, (1945): *How to solve it*, Princeton University Press, Princeton.

G. Polya, (1954): *Mathematics and Plausible Reasoning*, Princeton University Press, Princeton.

G. Polya, (1962): *Mathematical discovery*, John Wiley, New York.

M. M. Postnikov, (1960): *Teorija Galua*, GIFML, Moskva. English translation by L. Boron: *Fundamentals of Galois Theory*, Noordhoff, Groningen 1962.

C. J. Posy, ed. (1992): *Kant's Philosophy of Mathematics, Modern Essays*, Kluwer.

P. Rattansi, (1993): Newton und die Weisheit der Alten. in: Fauvel, ed. (1993), pp. 237–256.

B. Riemann, (1851): *Grundlagen für eine allgemeine Theorie der Funktionen einer veränderlichen complexen Grösse*, Inauguraldissertation Göttingen.

A. Robinson, (1961): Non-standard analysis. *Proc. Roy. Acad. Amsterdam*, **64**, pp. 432–440.

B. A. Rozenfeld, (1976): *Istorija neevklidovoj geometrii*, Nauka, Moskva. English translation by A. Shenitzer: *A history of non-Euclidean geometry: evolution of the concept of a geometric space*, Springer, New York 1988.

P. Rusnock, (2000): *Bolzano's philosophy and the emergence of modern mathematics*, Rodopi, Amsterdam.

B. Russell, (1897): *An Essay on the Foundations of Geometry*, Cambridge University Press.

B. Russell, (1902): Letter to Frege. in: van Heijenoort 1967, pp. 124–125.

B. Russell, (1908): Mathematical logic as based on the theory of types. *American journal of mathematics*, **30**, 222–262. In van Heijenoort 1967, pp. 150–182.

M. Schirn, ed. (1998): *The Philosophy of Mathematics Today*, Clarendon Press, Oxford.

E. Scholz, ed. (1990): *Geschichte der Algebra*, Wissenschaftsverlag, Mannheim.

J. Sebestik, (1992): *Logique et mathématique chez Bernard Bolzano*, VRIN Paris.

S. Shapiro, ed. (2005): *The Oxford Handbook of Philosophy of Mathematics and Logic*, Oxford University Press, New York.

P. A. Shirokov, (1955): *Kratkij očerk geometrii Lobačevskogo*, Nauka, Moskva. English translation by L. Boron: *A Sketch of the Fundamentals of Lobachevskian Geometry*, Noordhoff, Groningen 1964.

T. Skolem, (1922): Einige Bemerkungen zur axiomatischen Begründung der Mengenlehre. *Matema-tikerkongressen i Helsingfors*, Akademiska Bokhandeln, Helsinki 1923. English translation in van Heijenoort 1967, pp. 290–301.

M. Steiner, (1983): The philosophy of mathematics of Imre Lakatos, *The Journal of Philosophy*, **80**, pp. 502–521.

I. Stewart, (1989): *Galois theory*, Chapman and Hall, London.

R. N. Shcherbakov and L. Pichurin, (1979): *Ot proektivnoj geometrii k neevklidovoj*, (From the Projective Geometry to the non-Euclidean, in Russian.) Prosveščenije, Moskva.

A. Tarski, (1948): *A Decision Method for Elementary Algebra and Geometry*, University of California Pres, Berkeley 1951.

A. Tarski, (1959): What is elementary geometry? in: *The Axiomatic Method, with Special Reference to Geometry and Physics*, L. Henkin, P. Suppes, A. Tarski, eds., North Holland, Amsterdam, pp. 16–29

R. Torretti, (1978): *Philosophy of Geometry from Riemann to Poincaré*, D. Reidel, Dordrecht.

I. Toth, (1977): Geometria more ethico. in: *Naturwissenschaftliche Studien. Festschrift für Willi Hartner.* (Y. Maeyama and W. G. Saltzer, eds.), Wiesbaden, pp. 395–415.

B. L. van der Waerden, (1950): *Erwachende Wissenschaft*, Basel, Stuttgart 1966.

B. L. van der Waerden, (1985): *A History of Algebra, from al-Khwarizmí to Emmy Noether*, Springer Berlin.

J. van Heijenoort, (1967): *From Frege to Gödel, A Source Book in Mathematical Logic 1879–1931*, Harvard University Press, Cambridge, Ma.

J. von Neumann (1925): Eine Axiomatisierung der Mengenlehre. *Journal für reine und angewandte Mathematik*, **154**, 219–240. English translation in: van Heijenoort 1967, pp. 393–413.

F. Viéte, (1591): *Introduction to the Analytical Art*, in: Klein 1934, pp. 313–353.

L. da Vinci, (1492): Die Linearperspektive. in: *Sämtliche Gemälde und die Schriften zur Malerei*, A. Chastel, ed., Schirmer-Mosel, München 1990.

P. Vopenka, (1979): *Mathematics in the Alternative Set Theory*, Teubner, Leipzig.

P. Vopenka, (2000): *Úhelný kámen evropské vzdělanosti a moci.*(The Cornerstone of European Learning and Power, in Czech.) Práh, Praha.

J. Vuillemin, (1962): *La Philosophie de l'Algébre*, PUF, Paris.

H. Weber, (1893): Die allgemeinen Grundlagen der Galois'schen Gleichungstheorie. *Mathematische Annalen*, **43**, 521–549.

H. Weber, (1895): *Lehrbuch der Algebra*, Braunschweig.

L. Wittgenstein, (1921): *Tractatus Logico-philosophicus*, Suhrkamp, Frankfurt am Main 1989.

E. Zermelo, (1908): Untersuchungen über die Grundlagen der Mengenlehre. *Mathematische Annalen*, **65**, 261–281. English translation in: van Heijenoort 1967, pp. 199–215.

Y. Zheng, (1992): Non-Euclidean geometry and revolutions in mathematics. in: Gillies 1992, pp. 169–182.

Science Networks – Historical Studies (SNHS)

Edited by
Eberhard Knobloch, Technische Universität Berlin, Germany
Helge Kragh, University of Aarhus, Denmark
Erhard Scholz, Bergische Universität Wuppertal, Germany
In cooperation with an international editorial board

The publications in this series are limited to the fields of mathematics, physics, astronomy, and their applications. The publication language is preferentially English. The series is primarily designed to publish monographs. Annotated sources and exceptional biographies might be accepted in rare cases. The series is aimed primarily at historians of science and libraries; it should also appeal to interested specialists, students, and diploma and doctoral candidates. In cooperation with their international editorial board, the editors hope to place a unique publication at the disposal of science historians throughout the world.

SNHS 36: Kvasz, L.
Patterns of Change. Linguistic Innovations in the Development of Classical Mathematics (2008)
ISBN 978-3-7643-8839-3

SNHS 35: Caramalho Domingues, J.
Lacroix and the Calculus (2008)
ISBN 978-3-7643-8637-5

Silvestre François Lacroix (Paris, 1765 – ibid., 1843) was a most influential mathematical book author. His most famous work is the three-volume *Traité du calcul différentiel et du calcul intégral* — an encyclopedic appraisal of 18th-century calculus which remained the standard reference on the subject through much of the 19th century, in spite of Cauchy's reform of the subject in the 1820's.
Lacroix and the Calculus is the first major study of Lacroix's large *Traité*. It uses the unique and massive bibliography given by Lacroix to explore late 18th-century calculus, and the way it is reflected in Lacroix's account. Several particular aspects are addressed in detail, including: the foundations of differential calculus, analytic and differential geometry, conceptions of the integral, and types of solutions of differential equations (singular/complete/general integrals, geometrical interpretations, and generality of arbitrary functions).

SNHS 34: Høyrup, J.
Jacopo da Firenze's *Tractatus Algorismi* and Early Italian Abbacus Culture (2007)
ISBN 978-3-7643-8390-9

The abbacus books have long been supposed to be reduced versions of Leonardo Fibonacci's

Liber abbaci. Analysis of early abbacus books, not least of the first specimen treating of algebra — Jacopo da Firenze's *Tractatus algorismi* from 1307 — shows instead that abbacus mathematics was an exponent of a more widespread culture of commercial mathematics, already known by Fibonacci, and probably flourishing in Provence and/or Catalonia before it reached Italy. Abbacus algebra was inspired from a Romance-speaking region outside Italy, most likely located in the Provençal-Catalan area, and ultimately from a similar practitioners' level of Arabic mathematics.
The book contains, along with the English translation, an edition of Jacopo's *Tractatus* and a commentary analyzing Jacopo's mathematics and its links to Provençal, Catalan, Arabic, Indian and Latin medieval mathematics.

SNHS 33: De Risi, V.
Geometry and Monadology. Leibniz's *Analysis Situs* and Philosophy of Space (2007)
ISBN 978-3-7643-7985-8

SNHS 32: Krömer, R.
Tool and Object. A History and Philosophy of Category Theory (2007)
ISBN 978-3-7643-7523-2

SNHS 31: Keller, A.
Expounding the Mathematical Seed. Vol. 2: The Supplements (2006). ISBN 978-3-7643-7292-7

SNHS 30: Keller, A.
Expounding the Mathematical Seed. Vol. 1: The Translation (2006). ISBN 978-3-7643-7291-0

SNHS 30/31 Set: ISBN 978-3-7643-7299-6

BIRKHÄUSER

Science Networks – Historical Studies (SNHS)

Edited by
Eberhard Knobloch, Technische Universität Berlin, Germany
Helge Kragh, University of Aarhus, Denmark
Erhard Scholz, Bergische Universität Wuppertal, Germany

In cooperation with an international editorial board

SNHS 29: Guerraggio, A. / Nastasi, P.
Italian Mathematics Between the Two World
Wars (2005). ISBN 3-7643-6555-2

SNHS 28: Hesseling, D.
Gnomes in the Fog. The Reception of Brouwer's
Intuitionism in the 1920s (2003)
ISBN 3-7643-6536-6

SNHS 27: Dauben, J.W. / Scriba, C.J.
Writing the History of Mathematics – Its
Historical Development (2002)
ISBN 3-7643-6166-2 (Hardcover)
ISBN 3-7643-6167-0 (Softcover)

SNHS 26: Israel, G. / Millán Gasca, A.
The Biology of Numbers. The Correspondence
of Vito Volterra on Mathematical Biology (2002)
ISBN 978-3-7643-6514-1

SNHS 25: Siegmund-Schultze, R.
Rockefeller and the Internationalization of
Mathematics Between the Two World Wars.
(2001). ISBN 978-3-7643-6468-7

SNHS 24: Jensen, C.
Controversy and Consensus: Nuclear Beta
Decay 1911–1934. Edited by Aaserud, F., Kragh,
H., Rüdinger, E. and Stuewer, R. (2000)
ISBN 978-3-7643-5313-1

SNHS 23: Ferreirós, J.
Labyrinth of Thought. A History of Set Theory
and its Role in Modern Mathematics (2001)
ISBN 978-3-7643-5749-8

SNHS 22: Marage, P. / Wallenborn, G. (eds.)
The Solvay Councils and the Birth of Modern
Physics (1999). ISBN 978-3-7643-5705-4

SNHS 21: Sakarovitch, J.
Épures d'architecture. De la coupe des pierres à
la géométrie descriptive XVI–XIX siècles (1998)
ISBN 978-3-7643-5701-6

SNHS 20: Grattan-Guinness, I. / Bornet, G.
George Boole – Selected Manuscripts on Logic
and its Philosophy (1997)
ISBN 978-3-7643-5456-5

SNHS 19: Ullmann, D.
Chladni und die Entwicklung der Akustik
1750–1860 (1996). ISBN 978-3-7643-5398-8

SNHS 18: Hentschel, K. (ed.)
Physics and National Socialism. An Anthology of
Primary Sources (1996)
ISBN 978-3-7643-5312-4

SNHS 17: Corry, L.
Modern Algebra and the Rise of Mathematical
Structures (1996). ISBN 978-3-7643-5311-7

SNHS 16: Yavetz, I.
From Obscurity to Enigma. The Work of Oliver
Heaviside, 1872–1889 (1995)
ISBN 978-3-7643-5180-9

**SNHS 15: Sasaki, Ch. / Sugiura, M. / Dauben,
J.W.**, The Intersection of History and
Mathematics (1994). ISBN 978-3-7643-5029-1

SNHS 14: Klein, U.
Verbindung und Affinität. Die Grundlegung der
neuzeitlichen Chemie an der Wende vom 17.
zum 18. Jahrhundert (1994)
ISBN 978-3-7643-5003-1

SNHS 13: Vizgin, V.P.
Unified Field Theories in the first third of the 20th
Century (1994). ISBN 978-3-7643-2679-1

SNHS 12: Gorelik, G.E. / Frenkel, V.Y.
Matvei Petrovich Bronstein and the Soviet
Theoretical Physics in the Thirties (1994)
ISBN 978-3-7643-2752-1

SNHS 11: Reich, K.
Die Entwicklung des Tensorkalküls vom
absoluten Differentialkalkül zur
Relativitätstheorie (1994)
ISBN 978-3-7643-2814-6